EVASIVE WILDERNESS SURVIVAL TECHNIQUES

HOW TO SURVIVE IN THE WILD WHILE EVADING YOUR CAPTORS

SAM FURY

Illustrated by
NEIL GERMIO

NONFICTION SF BOOKS

WARNINGS AND DISCLAIMERS

The information in this publication is made public for reference only.

Neither the author, publisher, nor anyone else involved in the production of this publication is responsible for how the reader uses the information or the result of his/her actions.

CONTENTS

THANKS FOR YOUR PURCHASE

Did you know you can get FREE chapters of any SF Nonfiction Book you want?

https://offers.SFNonfictionBooks.com/Free-Chapters

You will also be among the first to know of FREE review copies, discount offers, bonus content, and more.

Go to:

https://offers.SFNonfictionBooks.com/Free-Chapters

Thanks again for your support.

INTRODUCTION

Evasive wilderness survival is the ability to keep yourself alive in a wilderness setting while avoiding capture (or recapture) by your enemy. An example scenario of this may be escaping a hostage situation where you were held captive in the wilderness.

This is the hardest type of wilderness survival there is, and the best type of survival to learn.

Evasive wilderness survival training focuses on worst-case scenarios, but is easily adapted to general wilderness survival. In fact, if you can survive in the wilderness under evasive circumstances, then non-evasive survival situations become much easier.

There are three key elements for succeeding in evasive wilderness survival:

- **Minimalism.** When you escape your "prison," you will probably not be able to take much with you.
- **Evasion.** Staying hidden from your enemy so you can avoid recapture.
- **Staying on the go.** Surviving while continuously moving away from your enemy and towards friendly territory. Your objective is to get to safety ASAP.

This no-fluff manual contains all the information you need to evade and survive in any terrain or climate, whether it be jungle, desert, arctic, etc. Even if these are the only wilderness survival lessons you learn, you will be very well equipped.

Training

Few people will carry a survival guide with them. It's better to learn the information (through practical activity where possible) so that you have the knowledge and hard skills.

You can use any single chapter from this book as a practical activity and/or a theory lesson, either individually, in succession as presented, or compiled into a multi-day survival course.

When training in this subject, please observe eco-friendly practices. Do not cut down live trees or kill animals. These things are okay in real survival situations, but not in training.

Local Knowledge

Every place in the world is different. You can adapt many survival skills (like the ones taught in this book) to a variety of situations, but having specific knowledge of the area you're in will make survival easier. Research the specific areas you're commonly in and/or plan to go to. Find out about the animals, useful plants, weather, etc.

Survival Needs

As an evasive survivor, you have needs that are the same as those of any wilderness survivor—namely food, water, shelter/warmth (clothing), fire, rescue, self-defense, first aid, and navigation.

However, you must also consider stealth while obtaining these things.

The specific things you need to acquire first depend on your situation, but the rule of three will be a consideration.

You can survive for three:

- Seconds without blood.
- Minutes without air.
- Hours without shelter.
- Days without water.
- Weeks without food.
- Months without human company.

The Will to Live

A big part of survival is maintaining your will to live and a strong belief that you will survive. Remember your reasons for living (e.g., loved ones) and have faith in yourself, your abilities, and your god if you have one. No matter what happens, do not give up your will to live, and always be prepared to seize opportunities.

RESOURCES AND IMPROVISED TOOLS

Obtaining a few items before you escape will make wilderness survival easier, but too much stuff becomes a burden. You also need to consider evading your enemy. Anything that will make noise or reflect light while you move will give you away.

Gather things that will help you meet your survival needs. Here's a list of those needs, with related items in parentheses:

- Shelter (winter clothing, poncho, cordage).
- Water (purification tablets, hiking filter).
- Food (fishing tackle, candy bars, foraging guide).
- Fire (matches, lighter, ferro rod).
- Medicine (first aid kit).
- Rescue (mirror, whistle, flashlight).
- Navigation (map, compass).
- Self-defense (knife, gun, club).

Your ability to gather survival resources before escaping may be minimal, but there will be additional opportunities once you're on the run. Always be on the lookout for useful items. A broken-down vehicle, for example, can provide cordage (wiring), fire (battery), digging tools (hubcaps), signal mirrors, and more.

Whatever resources you do have, ration them from the start. Even if you expect a quick rescue, things can go wrong. When you first escape, it's better to consume rations than to spend time scavenging. Once you have enough distance, live off the land as much as you can and conserve any excess stores for as long as possible.

KNIVES

A good knife is arguably the most useful survival tool there is.

If you have a choice, choose a carbon-steel blade with a V grind (Scandi grind). They are good all-around survival knives, and are easier to sharpen with improvised abrasives than knives made from other materials.

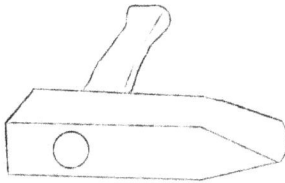

In an evasive survival situation, your chances of obtaining a real knife are slim. Here are some ways you can improvise one. If your enemy is close, consider the noise you will make constructing these before you do so.

Glass, Plastic, and Metal

You can turn hard plastic or soft metal into a blade by heating it up and hammering it into shape between two rocks before it cools. Sharpen the edge. Glass will already be sharp, but you can sharpen it more.

Bone

The larger the bone, the larger the knife you can make. Clean it well first.

Find a large, flat, hard rock for a table. You also need a hard, medium-sized stone with a round surface. This is your hammer stone.

Put the bone on the table and use your hammer stone to shatter it. Choose the best fragment to use as a knife. Ideally it will be one piece, with a sharp edge section and a handle section. Sharpen the edge more if you need to.

Stone

Stone blades are good for puncturing and chopping, but most won't hold a fine edge for long. Some exceptions are chert or flint.

First, look for a stone that already has a sharp edge. If you can't find one, making one isn't too hard, provided you can find the right stones.

To make a stone blade, you need two stones. The first is your blade stone. The bigger your blade stone, the easier the knife is to make. It also means you will get a bigger blade.

Look for a stone with a glassy surface. Check near rivers and creeks. If you find two of them, they should make a ringing, glass-like sound when hit together. Chert, flint, obsidian, and quartz are good examples.

The second stone you need is a hammer stone. Look for a hard, medium-sized round stone.

Place the blade stone on a larger rock, or on your thigh, and hold it firmly in place. Smack the hammer stone down on the edge of the blade stone, but don't do it too hard. Follow through with a strong, glancing blow.

When you do this correctly, blades will chip off the bottom of the blade stone. Sharpen their edges more if you need to.

Wooden Knife Handle

You can use improvised blades as they are, or make handles from wood.

To do the latter, split a piece of hard wood, insert your blade, and tie it in.

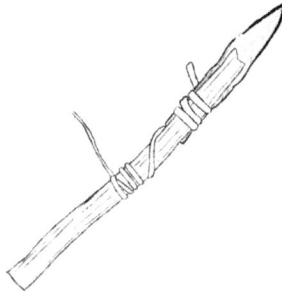

Related Chapters:

- Blade-Sharpening

BLADE-SHARPENING

Blade sharpening requires skill. Doing it incorrectly will do more damage to your blade than good. This chapter will teach you the correct ways to do it.

It's all about maintaining the V shape of your knife's edge. Ideally, you want it sharp enough to slice a piece of paper.

The following methods work best with real knives, but you can adapt them for improvised knives too.

Strop

Stropping keeps a sharp knife sharp, which is easier than sharpening a blunt knife. If you already have a sharp blade, strop it regularly by running the edge against a mild abrasive, such as a leather belt or thick cardboard.

Here is how to strop with a leather belt. Adapt it for whatever other material you want to use.

- Anchor the belt to your pants by the buckle.
- Hold the belt out tight in front of you with one hand, and hold the knife in the other.
- Place the blade flat on the belt, with the sharp edge facing you.
- Raise the back of the knife until one side of the V edge is flat on the belt (about a 20-degree angle). If you have light overhead, it is at the right angle when the shadow under the edge disappears.
- Apply slight downward pressure while keeping this angle.
- Scrape the knife back across the belt as you move it from the handle to the tip, then flip it over so the sharp edge faces away from you.
- Keep the same angle on the V edge and scrape the knife back towards you in the same manner.

- Repeat this back and forth.

The scraping is a slight diagonal motion, moving up/down and slicing at the same time.

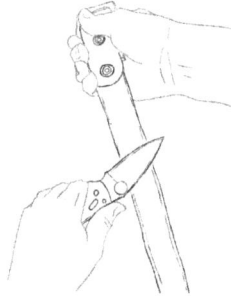

Steel

When the blade edge is a little dull, use steel to bring it back to sharpness.

A common way to do this in the wilderness is with a second steel knife. It needs to have a longitudinal groove pattern in the spine. If it doesn't, create one with fine sandpaper.

Use light pressure and the same 20-degree angle as with stropping. Scrape into the blade (the opposite of stropping) from the handle to the tip. Alternate sides with each stroke.

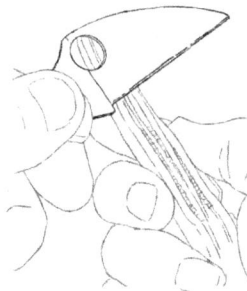

Steeling an edge is more abrasive than a strop, but you can use either method interchangeably if you only have one or the other.

Grinding

Use grinding when your blade is too dull to steel or strop—that is, when the V in the blade is more like a U.

Smooth, flat river rocks are good improvised sharpening stones. Rubbing two together will make one smoother. Use the rough side to remove burrs and the smooth side to get a fine edge. Use light-colored stones to make it easier to see your progress. To grind your blade:

- Wet the stone.
- Place blade on the stone at a 20-degree angle, with the sharp end facing away from you.
- With your fingers on top of the blade, move it clockwise on the stone.
- Apply steady pressure with your fingertips as the blade moves away from you, and release the pressure as you bring it back towards you.
- Keep the angle consistent and continue to wet the stone as needed.

Once all the burrs are gone on one side, turn the knife over. Use the same technique, but move counter-clockwise. Next, switch to the smooth side of the stone and use the same technique. Reduce pressure to get a finer edge, and finish up with steeling and/or stropping.

Other improvised grinding edges include the top of a car door window or any rough ceramic edge, such as the bottom of a mug.

Related Chapters:

- Knives

CLUBS

A club is any strong piece of wood you can find that has one heavier end. Use it as a weapon or a tool (for digging or to move things in a fire, for example). A good club is:

- Seasoned hardwood. There should be no moisture or green tint when you scrape the bark.
- Thin enough to hold comfortably, but thick enough so it won't break easily on impact.
- Long enough to do damage, but short enough to swing easily.

If you find one with a curve in it (like a boomerang) that's about 1/2m (1.5ft) long, you can use it to throw at and kill/injure small animals. This is a "stout" or "rabbit" stick.

Shaping one end into a flat head makes it a better digging tool. Loosen the soil with it, then use your hands or a flat rock to scoop the soil out.

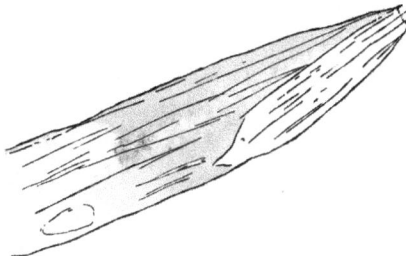

CORD

Cord (rope, string, etc.) is extremely useful, and easy to make out of things like fabric, fishing line, and shoelaces.

When you are fortunate enough to have some cord or any other type of material, avoid cutting it. Fold it instead, if possible. It is more versatile in larger pieces.

When there is no other material available (or you aren't willing to sacrifice it), then you can make cord out of other things including:

- Animal hair.
- Inner bark (e.g., cedar, chestnut, elm, hickory, linden, mulberry, or white oak). Shred the plant fibers from the inner bark.
- Fibrous stems (e.g., honeysuckle or stinging nettles).
- Grasses.
- Palms.
- Rushes.
- Sinew (dry tendons of large game).
- Rawhide.
- Vines. You can use strong vines without any other preparation, but plant fibers spun together are more durable.

Making Cord from Plant Material

When you think you have a suitable plant material, see if it can withstand the following tests. Soften stiff fibers first by soaking them in water. Then:

- Pull the ends in opposite directions.
- Twist and roll it between your fingers.
- Tie an overhand knot.

To turn the material into cord, twine it together.

The amount of material you need depends on how thick you want your cord to be. Divide it in half and rotate one half before recombining them. This will ensure an even consistency in your rope. Knot the material together at one end.

Divide the remaining side of the bundle into two even sections, and twist them both clockwise to create two strands. Next, twist one of the strands around the other in a counter-clockwise direction. Tie the end to prevent it unraveling.

You can join shorter lengths together by splicing them. Do this by twisting the ends of their strands together while they are in two strands (that is, before the counter-clockwise twisting). Twist one small bunch on each side of each of the strands and then continue to twist as before. You can do this as much as you want until you get the length of rope you need.

Make thicker ropes by using larger bundles of grass or by twisting multiple ropes together.

Making Cord from Animals

In a survival situation, you may be fortunate enough to capture game. Waste nothing.

Sinew is an excellent material for small lashings. Remove the tendons from game animals and dry them. Once they are completely dry, hammer them until they are fibrous. Add some moisture so you can twist the fibers together. You could also braid them together, which will make the product stronger. Sinew is sticky when wet and hardens when dry. Lash small items together while the sinew is wet; since it dries hard, you won't need to use knots.

When the job is too big for sinew, use rawhide. Skin any medium to large game and clean the skin very well. There must be no fat or meat on it, though hair or fur is okay. Dry it completely. If there are folds that will capture moisture, stretch the skin out. Once it's dry, cut it into a continuous 10mm wide length (1/2in). The best way to do this is to begin in the middle of the skin and cut outwards in a circle, expanding the spiral as you go. To use the rawhide, soak it until it's soft, which usually takes several hours. Use it wet, and stretch it as much as you can as you do so. Leave it to dry.

The information in this chapter was from the book *Emergency Roping and Bouldering*:

www.SFNonfictionBooks.com/Emergency-Roping-Bouldering

STEALTH MOVEMENT

In this section you'll discover how to leave minimal signs of presence while evading your enemy. You'll also learn navigation, safe ways to move in various terrains, and more.

OBSERVATION

Constant observation using all your senses is required when you're moving. Even when you stop, you must keep observing. Observe your enemy and/or any obstacles in your way, so you can choose how and when to move.

Searching Ground

Use this method to look for signs of your enemy, or anything else you want to look for, from a stationary position. It will help if you have something specific to look for (certain equipment, humans, dogs, vehicles, etc.).

Divide the ground into three ranges: immediate, medium, and long. Scan each section from right to left. Start with the immediate range, and work your way back systematically.

Right to left is better than left to right because we read from left to right and are more likely to overlook things if we follow that habit. Horizontal scanning is better than vertical, as that way you don't have to be continuously adjusting for distance and scale.

When you come across areas that are more likely to hide something, take a bit more time to search and look for parts of objects as well as whole ones. Things may be hidden behind something, but with bits of them still visible.

Look through visual screens, e.g., vegetation. If you want to look further, make a small head movement.

Tips for Seeing in the Dark

It takes 30 minutes for your eyes to fully adjust to the dark (night vision) and you need at least a little ambient light from a source like the moon.

Once your eyes have adjusted to the dark, you need to protect them. A flash of light can ruin your night vision in a second. When there is a bright area you want to observe, cover one eye to preserve it while you use the other one to look.

Even with your night vision, objects in the dark are harder to make out. Looking next to them will make them clearer. Changing your focal point every few seconds (up, down, to the sides) will also help.

Things may seem to move. Make sure they're staying still with the sticky finger method. Stretch a finger out in front of you and "stick" an object to it.

When you need extra light to see (if you're reading a map, for example), use red or blue light. It does minimal damage to your night vision and is harder for your enemy to spot. Don't rely solely on your vision. Sound, smell, and touch can tell you many things.

Hearing is a human's next best sense, and you can often hear things that are out of sight. Stay still, open your mouth a little, and turn your ear in the direction you want to hear.

The wind can carry smells quite far, and some smells, like food cooking or smoke, are very distinctive to humans. Turn your nose up toward the wind and smell like a dog does, taking many small sniffs. Concentrate on the inside of your nose and try to determine what the smell is.

When you can't see anything at all, it's safer to stay still until there's light, but certain circumstances may require you to move. In this case, you need to feel your way around. Move slowly, testing every movement.

Lift your feet high to give yourself the best chance of clearing any obstacles, but ensure you do not lose balance. Stretch your hands in front of you to feel for obstacles. Use the back of your hand to feel stuff, in case it's sharp or hot. This will protect the inside of your hand and the arteries in your arm.

COVER AND CONCEALMENT

Cover and concealment are different. Both are useful for stealth.

Concealment is anything between you and your enemy that hides you from sight. Vegetation is a good example of concealment. The more of it there is between you and your enemy, the harder it will be for him to see you.

Cover will hide you from sight too, but will also stop bullets. Many solid objects do not qualify as cover. Bullets will go straight through wooden fences, car doors, windows, etc.

Solid concrete, thick metal, depressions in the earth, and large trees have a much better chance of providing you cover. The more powerful the gun (or blast), the thicker the cover needs to be.

If your enemy is trying to shoot you, seek cover. If he just wants to find you, concealment is enough.

When covering ground, move from cover (or concealment) to cover, stopping at each one to observe. Make sure you know your next place of cover or concealment before leaving your current one.

CAMOUFLAGE

Having a good understanding of the principles of camouflage will help you in all areas of stealth movement. Most of these things are intertwined. Use them together for the best results.

Shape

The human shape (or anything) is distinctive, but there are ways to distort it. For example, you can attach local vegetation to yourself or adjust your posture.

Size

The bigger things are, the easier they are to spot, and the harder they are to hide. You can make yourself smaller by getting lower to the ground and/or standing sideways for a slimmer profile.

Silhouette

When an object contrasts against a plain background, the shape of its outline is its silhouette. This is most prominent when there is a dark object on a light background, or vice versa. Examples of plain backgrounds in nature are the sky and the sea.

Even a slight shade difference is enough for a keen observer to spot a silhouette. For example, wearing black clothing creates more of a contrast at night than dark blue clothing does.

To minimize your silhouette, keep to low ground and/or lower your physical profile.

Color and Texture

Every environment has certain colors and textures, and if you don't mimic those, you'll stand out.

Contrasting colors, like light-colored hair in the forest or black clothing in the snow, stand out more.

Textures may be rocky, leafy, smooth, etc.

Distort your color and texture and that of your equipment with things like mud, vegetation, charcoal, or cloth. Consider depth of features. Use lighter colors on shaded areas (around the eyes and under the chin) and darker colors on features that stick out more (forehead, nose, cheekbones, chin, and ears).

When using vegetation to blend in, ensure its color and texture continue to match the terrain as you move, since the vegetation will change and the leaves will wilt.

When you need to hide yourself quickly, lay down flat and cover yourself with foliage.

Shine and Reflection

Shine is anything that reflects light, including oily skin. An enemy can spot shine from great distances if the angle of light is correct.

Cover glass, metal, and anything else that shines (zips, buckles, jewelry, watch faces, etc.), no matter how small it is. If you need to wear glasses, line the outsides of the lenses with a thin layer of dust to reduce the reflection of light.

Reflection isn't a big deal at a distance, but can give you away if you're careless. Avoid mirrors, glass, and anything that gives a reflection. Stay outside the field of reflection—by crouching under mirrors, for example.

Light and Shadow

Avoid moving in and using light to see as much as you can, especially at night time.

Moving under or near light makes you more visible and casts your shadow. This can give you away even when the rest of you is hidden.

Always be aware of where your shadow falls, and keep in mind that the direction of the shadow will shift with the movement of the sun or other changes in light.

Knock out lights (trip fuses or break globes) if doing so won't give away your position.

The outer edges of the shadows are lighter and the deeper parts are darker. Keep in darker parts of the shadow when possible.

Your silhouette may still be seen against lighter shadows, so keep low and still until you need to move.

If you must use a flashlight, cover the head of it with your hand. If possible, use a colored lens filter.

Noise

When you're close to your enemy, you must be careful the noise you ma. The slower you move, the quieter you can be.

Ensure there's nothing on you that will rattle, jingle, vibrate, ring, or chime. If possible, jump up and down and listen for any noise you make, and secure anything you need to.

When you have the choice, keep to quieter surfaces, such as bare earth, flat concrete, wet leaves, and large rocks.

Time your movement to coincide with ambient sounds (passing traffic, barking dogs, rain, or gusts of wind) conceal yourself.

If you hear a noise that might be your enemy, freeze and observe. Get to the ground or behind cover if you can do it without getting spotted.

Use noise and movement to distract an opponent. For example, throw something in the opposite direction from where you want to go, so your enemy's attention will focus on it.

Put down small objects by touching your hand to the surface first, then lowering the object down.

Scent

Humans have certain smells (soap, food, body odor). Do the following to lessen your scent:

- Wash yourself and your clothes without using soap.
- Avoiding-strong smelling foods like those with garlic and spices.
- Don't use anything that smells unnatural, such as cologne, tobacco, or gum.
- Rub your clothes in aromatic plants (pine needles, for example) taken from your surroundings.

Pay attention if you smell the signs of humans, such as fire, gasoline, or cooking.

Keep downwind of your enemy when possible, especially if they're using dogs.

Related Chapters:

- Clothing

MODES OF MOVEMENT

When evading your enemy, you need to compromise between stealth and speed. What you choose depends on your circumstance, but in general, the closer you are to your enemy, the stealthier you need to be.

For maximum stealth, move low and slow. The lower you are, the "smaller" you are, and the harder you are to see.

The slower you are, the less likely you are to attract the eye and the less noise you make.

When the enemy is close, go as low and slow as you can. If he looks in your direction, freeze. You can move faster as you get further away.

There are four basic ways you can move when you're on foot.

Walk

Walking is a good compromise between speed and stealth. You can control your speed depending on your needs, and easily shift from walking into other positions, like breaking into a run or crouching down.

The basic principles of stealth-walking apply to all types of movement.

To walk as quietly as possible, place all your weight on one foot and lift your other foot high enough to clear any obstacles, but not so high that you lose your balance. Small steps are easier to control.

Test the ground by carefully pressing down on it with the outside edge of the ball of your lead foot. If the step is going to make noise —if you're stepping on a twig, for example—test a different area. On loose ground, such as that covered with leaves, you can place your feet under the foliage.

When you find a quiet spot and are ready to continue, roll to the inside ball of your foot and then to your heel, and finally to your toes. Shift your weight to your lead foot, ensure you're balanced, and repeat the process with your rear leg.

On hard ground that is noisy, muscle control becomes paramount. The slower you go, the more control you have over your muscles and the quieter you can be. You want to be able to stop at any stage of the movement and hold your position for as long as you need to.

Keep your arms and hands close to your body, ensuring they don't hit anything.

As you walk in this manner, use relaxed, normal breathing. It encourages naturalness of movement and helps to prevent gasping if you misstep or lose balance.

Wrap cloth around your feet to muffle sounds if possible.

Stomach Crawl

This is the stealthiest way to move because you have the lowest profile.

Do not slide on your stomach. That leaves too much of a trail and makes noise. Instead, use your hands and toes to do a pushup that moves your body forward. Lower yourself to the ground, move your hands up to the pushup position again, and repeat the movement.

Crawl

When crawling on your hands and knees, test the ground with your hands before applying your weight. Put your knees in the exact same place your hands went.

Run

Running while crouching is a good way to cover short distances while no one is watching. Use this technique to get past a guard whose back is momentarily turned, for example.

Going into a full run is not at all stealthy, but it is the fastest way to create distance, which is important for evasion. As soon as you're confident you are out of sight, or if you've definitely been spotted, break into a full run.

EVADE TRACKERS

Once you're on the run, always assume you are being hunted, and act accordingly until you are 100% safe.

To evade trackers, you'll need to:

- Put as much distance (and therefore time) between you and the tracker(s) as possible.
- Leave minimal signs of presence.
- Create false signs and other obstacles to confuse/slow down the tracker(s).

Minimize Signs of Presence

A sign of presence is any disturbance you make to the natural surroundings. Not leaving any signs of presence is near impossible, especially when you're moving quickly, but here are some tips for minimizing them:

- Avoid touching your surroundings as much as you can. Try not to grab shrubs, lean on trees, or break spiders' webs, for example
- Be careful not to break small branches in your way. Instead, bend them. If that's not possible, go under, over, or around them.
- Be careful of leaving scuff marks when you climb over things.
- Be wary of transferring one type of ground to another (sand or water onto rocks, for instance).
- Do not leave any litter.
- If your clothing gets caught on something, ensure no part of it is left behind.
- Move during bad weather, such as strong winds, rain, or snow.
- Tiptoe through soft ground to minimize your tracks.

- Walk in existing footprints.
- Walk on hard surfaces like rock where possible, so you leave less of a footprint.
- Walk on the inside of your foot to avoid leaving a heel or toe mark.
- If you can, wrap your shoes (in cloth, duct tape, etc.) to reduce your footprint, but be careful not to leave traces of the cloth.

Evading Dogs

If your tracker has a dog, you must take extra care to mask your scent and sound:

- Avoid strong-smelling substances like smoke or animal scat. Hide contaminated clothing under rocks in a stream, or bury it if that is not possible.
- Cross still water diagonally.
- Don't make fresh tracks. Use existing trails or move just to the side of animal tracks.
- Go through dusty, polluted, or animal-filled or other environments that will confuse the dog's sense of smell.
- In heavy foliage and animal fields, use an erratic path by changing direction frequently.
- Keep going. Tire the dog out and it will make mistakes.
- Like many animals, dogs smell fear. Keeping your cool will make you less "stinky."
- Separate the handler from the dog.
- Stay hidden. Although a dog's sight isn't that great, they will see movement and use their other senses to follow up.
- Use a vehicle.
- Use terrain that holds less scent, such as water, hard rock, metal, ice, or sand. Cross and re-cross these surfaces at intervals to make false trails while not hindering your speed too much. Walking exclusively on one, such as water, is slow.

- Walk in running water for a little bit and exit where your footprint will not show.

Taking out a dog is good, but taking out the handler is better.

False Signs, Obstacles, and Deceptive Tracks

False signs, obstacles, and deceptive tracks take time to make, but they are worth it if you successfully fool your tracker. Experienced trackers will be hard to trick, but most people have little experience. Here are some ways to stall them:

- Alter the length of your stride.
- Change your shoes so you leave a different tread mark.
- If you're in a group, split up and meet up later, even if it's just in a few hundred meters or so.
- Leave signs of presence in one place and hide in another, ready to ambush your tracker.
- Open a door or gate as you pass so he thinks you went through it.
- Set up traps. Even the illusion of traps will slow your tracker down in fear for his safety.
- Use a stick to bend grass and branches in a different direction than the one in which you're going.
- Use deceptive tracks.
- Walk in reverse or tie your shoes on backward.

Deceptive Turn

When you want to turn someone away from your current direction, use a deceptive turn. It works best when your deceptive path leads to an area that's hard to track in, such as one with water or hard surfaces, as it will take him longer to figure out.

Walk 5m (16ft) or so past your turnoff and leave a sign of presence to make your tracker think that this is the way you went. You want your tracker to notice it, but not to know you did it on purpose, so don't make it too obvious.

After leaving your sign of presence, walk backwards to your turnoff point and then head in your new direction. Be careful not to leave signs of presence while walking backward or when heading in your new direction.

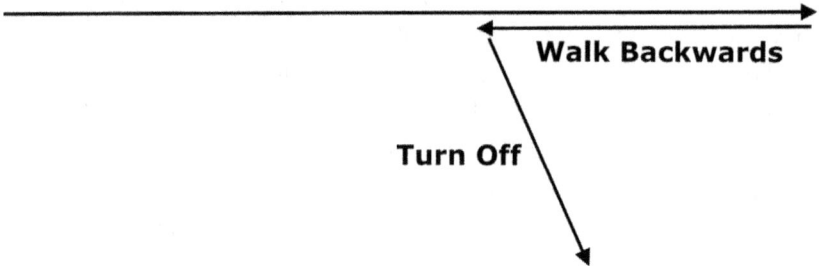

Walk Backwards

Turn Off

Trail Deception

When you see a trail up ahead, use this deceptive track to make your tracker think you're using the trail instead of sticking to your original path.

Approach the trail at a 45-degree angle. Do it from about 100m (330ft) out if possible.

Walk along the trail for 25m (80ft) or so. Leave some signs of presence, but don't make them obvious, and then walk backwards to where you entered the trail. Cross the trail and walk away from it in a 45-degree angle until you return to your original line of travel. Be careful not to leave any signs of presence as you do this.

You can also do this in a stream.

Trail

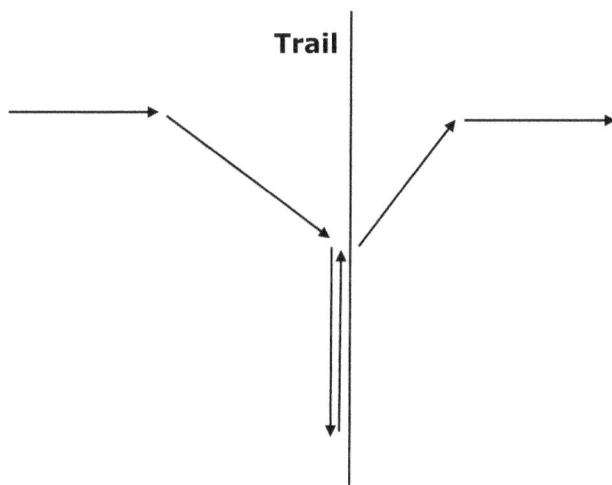

Stream Exit Deception

Walking in a stream is a good way to evade trackers and dogs, but eventually you'll need to exit it, and the transference of water is a big sign of presence.

To combat that, walk out of the stream for about 30m (100ft) every now and again. Then walk backwards along the same path and continue to walk in the stream. Make the false exits at least 100m (330ft) apart, and use a different exit direction each time.

When you leave the stream for real, choose a spot with rocks, roots, or other features that will ensure you'll leave minimal signs of presence.

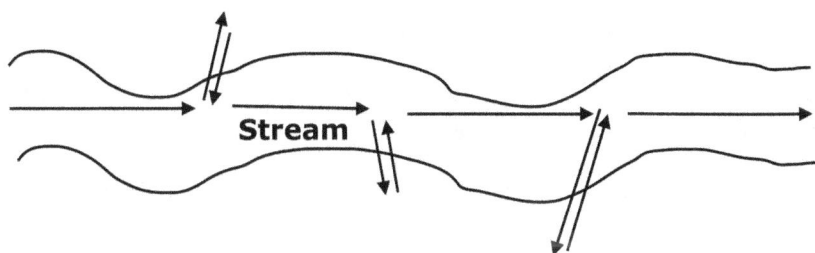

Stream

MAP AND COMPASS NAVIGATION

The ability to navigate is a useful and potentially lifesaving skill. In this chapter, you'll learn how to use a map and/or a compass to stay on your intended path and/or to figure out where you are.

GPS is good, but a map and compass are more reliable. Batteries can run out and satellite signals can be iffy. If you do have a GPS, use it to confirm your position when you're unsure. Only turn it on where it will have good signal, such as on open and/or high ground, and keep it off the rest of the time to conserve the batteries.

Maps

Do your best to get hold of a map while you're in captivity so you can plan your escape better. If you can't steal one, draw it on the inside of your clothing. Once you're out, improve your homemade map by getting to high ground and surveying the area.

There are two basic types of maps:

- Topographic (topo) maps are the more complicated-looking ones with detailed markings.
- Planimetric maps are like street or tourist maps.

If you can use a topographic map, then you can use a planimetric one, but that's not necessarily true the other way around. Topo maps are also more accurate. All of them are slightly different, but they all will have at least some of the following features:

Compass rose. The compass rose shows north on the map. Grid lines are usually not aligned with north.

Contour lines. Contour lines are the "squiggly circles." They tell you the altitude of the land above sea level, which indicates the steepness of the landscape. The closer the contour lines, the steeper the slope.

The number inside the contour line represents the number of meters above sea level. Not every contour line will have a number, but they are all the same distance apart. Every contour line might represent another 10 meters of altitude, for example. A dashed line means there is no rise in elevation.

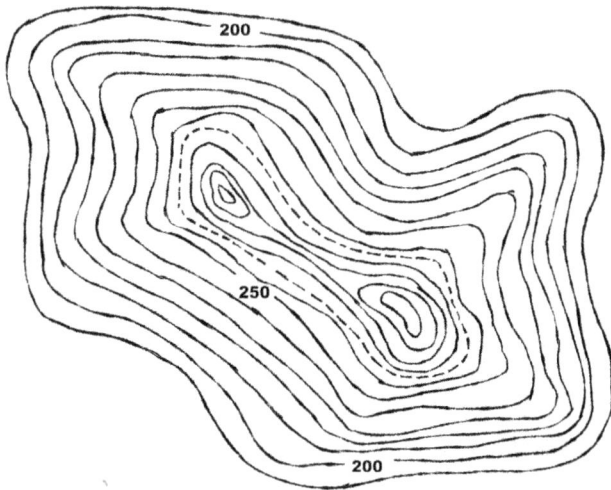

Coordinates. The map coordinates are the numbers that run along the edges of the pages. They usually represent latitude and longitude, and are a good way to communicate position.

When sharing co-ordinates, give the horizontal number first, then the vertical. To remember this, recall that you must walk on the ground before climbing a tree.

Key/legend. This tells you what each symbol on the map represents. In general:

- Blue = Water
- Green = Vegetation
- Black = Manmade structures

Magnetic variation/declination. Magnetic variation is important for you to be able to accurately align your compass with the location the map is based on. It is the angle between magnetic north and true north.

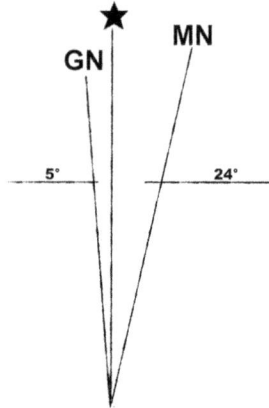

Scale. Scale is the size of the map in relation to real life. Knowing this is useful for judging distance. For example, if the scale is 1:20,000, then the map is 20,000 times smaller than reality. The widths of lines (streams, roads, etc.) are usually not to scale, but the lengths are.

SCALE 1:24 000

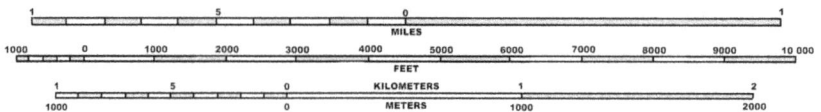

Land features. The ability to recognize land features on a map will help you choose the best way to get to where you want to go. It's also useful for finding and confirming your location in relation to the map.

There are many types of land features on a map, but for general navigation, you only need to know the main ones:

1. **Peak.** The highest point of high ground (such as the top of a hill).

2. **Ridgeline**. The line of high ground (from peak to peak, for example).
3. **Ridge**. The line of sloping ground (such as the side of a hill).
4. **Saddle**. The low point between two areas of high ground (such as the low point of a ridge line).
5. **Valley**. The line of low ground (between two hills, where water runs, for instance).

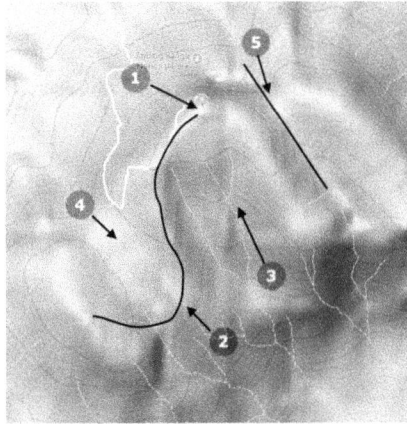

Compasses

Compasses come in many sizes. Larger ones usually have more features, but a small one is easier to conceal and will do the job if it's a good-quality product like a Silva or Suunto. Don't skimp on a good compass. An unreliable compass is worse than no compass at all.

Every compass has the following components:

Baseplate. This is the flat, hard surface. A clear base plate is preferable so you can see the map through it.

Heading arrow. This is the arrow at the front of the baseplate that shows your direction.

Housing. The housing contains liquid and the needle that faces north (the north needle).

Orienting arrow. An arrow marked inside the housing of the compass.

Index Pointer. A small line at the bottom of the heading arrow. Use it to read your bearing.

Rotating bezel/dial or Azimuth ring. This is the part you can spin. It is marked with the degrees. Get a compass with a rotating bezel marked in two-degree increments, preferably with north-south-east-west (NESW) points.

Scales. These are the ruler-like markings on the edge of the baseplate. Use them with maps that have different scales.

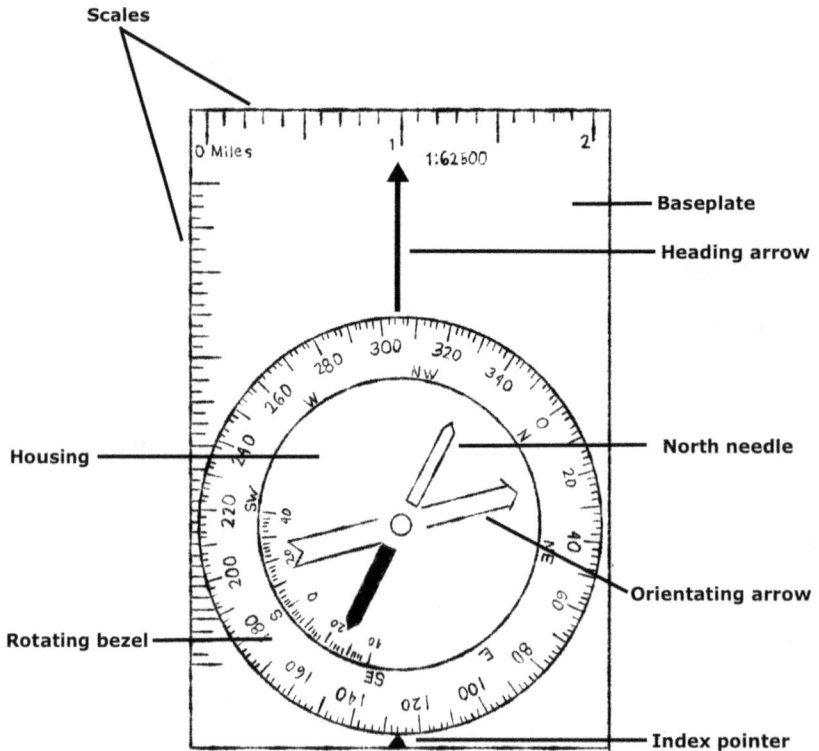

False Readings

At the magnetic poles (north and south) there is a lot of "pull" and your compass will not point north properly. Other things, like wristwatches, keys, cell phones, speakers, geological formations, magnetized rocks, and power lines, may affect a compass too. Use survival navigation to confirm your compass is pointing north.

Route Planning

Creating a route plan will help you keep track of where you're going. In a covert situation, do not mark the map or write out a route plan. Memorize it instead.

To create a detailed route plan, you need a pencil, paper, a compass, and a topographic map.

	Starting Co-ordinates						
	Destination Co-ordinates						
	Magnetic Variation						
Leg	Starting Waypoint (Label/Co-ordinates)	Destination Waypoint (Label/Co-ordinates)	Bearing	Distance	Elevation	Estimated Time	Notes
1							
2							
3							
4							
		Totals					

See the end of this chapter for a larger version of this route plan template.

Choosing the Route

When choosing a route, you must compromise between the easiest terrain and where your enemy will expect you to go. Plan to head for the nearest friendly settlement, but not directly to it. Consider the places where your tracker is likely to look for you, and avoid those areas.

If you don't know where the nearest civilization is, your best bet is to follow waterways downstream, but not too closely. They are obvious places for your enemy to look.

When there are numerous options for getting to safety, use one or two radical changes in direction. This forces your tracker to follow you instead of racing ahead and cutting you off.

Other things to consider are:

- Opportunities for gathering survival supplies along the way (food, water, shelter, etc.).
- The availability of cover and concealment.
- Rally points, if you're in a group.
- Avoiding obstacles. The shortest path is not necessarily the quickest. Going along a long road is safer than crossing mountainous terrain, for example.

In a non-covert situation, stay close to transport routes (air, ground, or water) and/or open ground, so it's easier to attract rescuers' attention. Once you have a primary route, choose some alternate ones as well. If you're in a group, consider splitting up.

Mark Your Map

Mark your starting point, destination, and routes on your map in pencil so you can erase them later.

Mark down waypoints as well. Waypoints are places where you can confirm your position in relation to the map, such as prominent features, direction changes, and places you are likely to make mistakes.

Record Information on Your Route Plan

Write the following on the top of your piece of paper (which is your route plan):

- Title, if you have multiple route plans.
- Coordinates of your starting point and destination.
- Magnetic variation.
- Waypoints. Use short labels and coordinates.

Take a Map Bearing

A bearing tells you the direction you need to follow. To take a bearing:

- Lay your compass flat on the map.
- Move it so the edge of your compass joins your starting position to your first waypoint (or waypoints 2 and 3, 3 and 4, etc.).
- If your compass edge is not big enough, draw a straight pencil line between the two points and align your compass with that line.
- Rotate the bezel until the compass's meridian lines match up with the north-south lines on the map—that is, so north is facing north.
- Read your bearing at the index pointer.
- Make the adjustment for magnetic variation and enter the compass bearing in your route plan.

Note: The bottom of the compass (180 degrees around) is your reverse bearing (from B to A).

Magnetic Variation/Declination

There are three types of north: grid north, true north, and magnetic north.

The difference between grid north and true north is negligible, so you don't have to worry about it, but the difference between grid/true north and magnetic north is something you need to compensate for.

When you take a map bearing, the direction you get is in reference to grid north, but your compass points to magnetic north. This difference between grid north and magnetic north is known as magnetic variation/declination. If you don't adjust for magnetic variation, which changes depending on time and place, you'll travel in the wrong direction.

The magnetic variation and the information needed to update it will be written on your map. If it isn't, you can visit:

http://www.ngdc.noaa.gov/geomag-web/#declination

Once you know the variation, subtract it from your grid bearing to get your magnetic bearing.

Example one:

- Your map bearing (grid/true) is 50°.
- The magnetic variation is 18°E, which is positive (+18°) because east is to the right of north.
- Subtract 18° from 50° to get your compass (magnetic) bearing of °32.

Example two:

- Your map bearing is 43°.
- The magnetic variation is 06°W, which is negative (-06°) because west is to the left of north, i.e.:
- 43° - (-06°) = 49°.
- 49° is your magnetic bearing, so that's what you'll set your compass to.

To convert a magnetic bearing to a true bearing, add the magnetic declination.

Example:

- You take a compass bearing of 30° and you want to plot it on the map.

- The magnetic variation is 10°W.
- 30° + (-10°) = 20°.
- 20° is your grid bearing, which you will plot on your map.

Calculate Distance

Measure the distance between two waypoints using a ruler or the scale on the side of your compass.

If the path is not straight, use a piece of string, such as your compass's lanyard, to trace the route. Straighten the string and measure the length as normal.

Another way to measure distance on a curved route is with a pencil/pen and paper:

- Align the straight edge of the paper with the space between your starting point and where the first bend in the route is.
- Make a small mark on the edge of the paper at the starting point.
- Make another mark on the paper where the bend is.
- Align the paper with the space between the first bend and the next one, so that your first bend mark is on the first bend.
- Mark the second bend.
- Repeat this process for all the bends.
- When you're finished, your piece of paper will have a little mark for each time there's a bend in your route between two waypoints.
- Measure from your first mark (starting point) to your last one (destination).

Once you have your measurement, convert it to actual distance using the scale.

Example one:

- If the map scale is 1:50,000, it means 1cm on the map is equal to 50,000cm in real life.
- A measured distance of 5cm is actually 250,000cm (5 x 50,000).
- 250,000cm = 2500m = 2.5km (100cm in a meter, 1000m in a kilometer).

Example two:

- If the map scale is 1:25,000, it means 1cm on the map is equal to 25,000cm in real life.
- A measured distance of 3.2cm is actually 80,000cm (25,000 x 3.2).
- 80,000cm = 800m.

Elevation

Count the contour lines to record the change in altitude between your waypoints (+5m, -10m, or 0, for example).

Estimate Timing

If you know the distance, terrain, and personal variables (fitness, load carried, etc.) you can estimate how long it will take you to get from waypoint to waypoint. Naismith's rule is a good starting point for this. The version given below is not the original Naismith's rule, but one that incorporates common variations/additions:

- Allow one hour to walk 4km (originally 5km).
- Add one hour for every 600m of (2,000ft) ascent.
- Subtract 10 minutes for every 300m (1000ft) of gentle descent.
- Add 10 minutes for every 300m (1000ft) of steep descent.

The above assumes the hiker is of average health, on typical terrain, and with no other complications or variables, such as rest stops, considered. Adjust it according to your actual circumstances and allow for rest times (10 minutes' rest after every hour of walking, for example). Remember that when you're in a group, you're only as fast as your slowest member.

Notes

Record any special notes, such as features you'll come across, to help with your navigation. As well, note any information that you come across along the way that you feel will be useful in the future.

Repeat and Total

Record bearings, distances, and timings for all the waypoints on your map. Calculate the total distance and estimated time.

Following Compass Bearings

Rotate your compass bezel so that the index pointer is on the bearing you wish to walk on (the one to your first waypoint, for example).

Hold your palm out flat with your fingers facing directly in front of you, and place the compass flat on your palm, with the heading arrow pointing down your middle finger.

Rotate your whole body (not just your hand) until the north side of the north needle is within the orienting arrow. Look ahead in that direction and choose a good landmark along that bearing that you can walk to. (**Note:** Landmarks may move in arctic or desert conditions.) Once you get to that object, reshoot your bearing.

Another way, which is more accurate, is to have a partner walk out along your bearing so he's just in your sights. Use hands signals (or radio) to move him exactly in line with your bearing. Move to your partner's position and then repeat the process.

When in featureless terrain, you can align yourself by looking back at your tracks to ensure they are in a straight line.

At night you can use the stars, but they move over time (due to the earth's rotation), so retake the bearing regularly.

If you go around an obstacle, be sure to get back to your bearing once you've passed it.

Map to Ground Orientation

Orienting a map means aligning it with what you see in real life. For example, if the river is to your left, turn the map so the river is to the left.

This is useful for seeing clearly where you are and/or in which direction you need to go.

Triangulation/Resection

Triangulation uses geometry to calculate your position on a map. To do it:

- Pick two or three features you can see and identify on your map, such as mountains. It's easier to do this from high ground.
- Take a bearing from you to the first feature.
- Use this bearing to draw a line on your map that passes through the feature. Don't forget to adjust for magnetic variation.
- Repeat this for the other one or two features.
- The point where the lines intersect is your approximate location. If you used three features, you'll be somewhere in the triangle the lines create. This is more accurate than a two-line intersection.
- Identify landmarks close to you to get an exact location.

You can also use triangulation to pinpoint a feature on your map. Here's how:

- From your current position, take a bearing to the feature.
- Draw a line on your map from your position along this bearing.
- Go to a second point that you can identify on your map and take another bearing to the feature.
- Draw a line on your map from this new position along the bearing.
- The point where the two lines intersect is the position of the feature.
- For increased accuracy, take a third bearing.

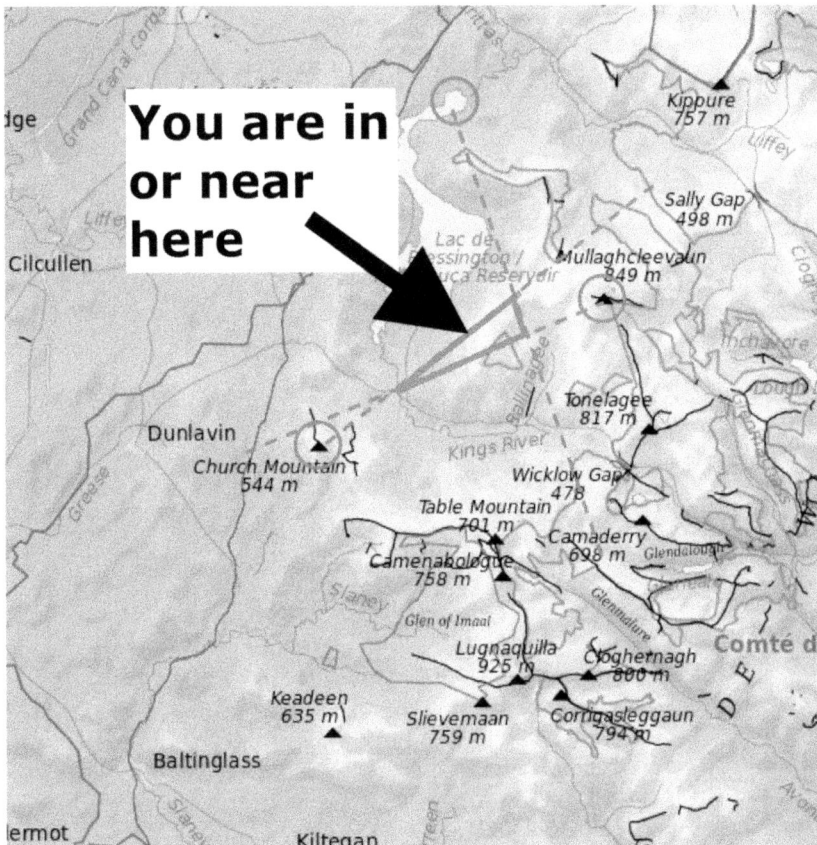

Starting Co-ordinates

Destination Co-ordinates

Magnetic Variation

Leg	Starting Waypoint (Label/Co-ordinates)	Destination Waypoint (Label/Co-ordinates)	Bearing	Distance	Elevation	Estimated Time	Notes
1							
2							
3							
4							
Totals							

Related Chapters:

- Cover & Concealment
- Survival Navigation

SURVIVAL NAVIGATION

Survival navigation is the ability to navigate without a map, compass, or GPS. None of these methods alone are very accurate, but if you combine several of them, the combination can be enough to get you to safety.

Even if you have a map, compass, and/or GPS, it's useful to confirm your direction with these survival navigation methods (to make sure your compass isn't giving you a false reading, for example).

Basic Sun Movement

The sun rises in the east and sets in the west (roughly). This is most accurate closest to the times of the equinoxes (March-April and September-October).

During the summer, the sun will rise and set a bit further north of east and west. In winter, the sun will rise and set a bit further south. This is true whether you're in the Southern or Northern Hemisphere, since the seasons are at different times of year. As well, the further you are from the equator, the further from east and west the sun will rise/set.

Shadow Stick

This method uses the shadows created by the sun or a bright moon to find north. It's less accurate near the equator or at the polar regions, where the shadows will be too short or long.

When the sun is at its highest point, shadows are at their smallest. If they're visible, they'll point north/south. This time of day is "local apparent noon," and is usually not 12:00 noon.

The shadow-stick method is most accurate when used within two hours of local apparent noon. Here's how to do it:

- Stick a straight stick into the ground. Try to find one that's 1m (3ft) long.
- Mark where the tip of its shadow is.
- Wait 10 minutes and mark the tip again.
- Do this a few more times.
- The markers form a west-to-east line. The first marker is west. Put your left foot on the first marker and your right foot on the last one.
- When you're in the Northern Hemisphere, you'll face north. If you're in the Southern Hemisphere, you'll face south.

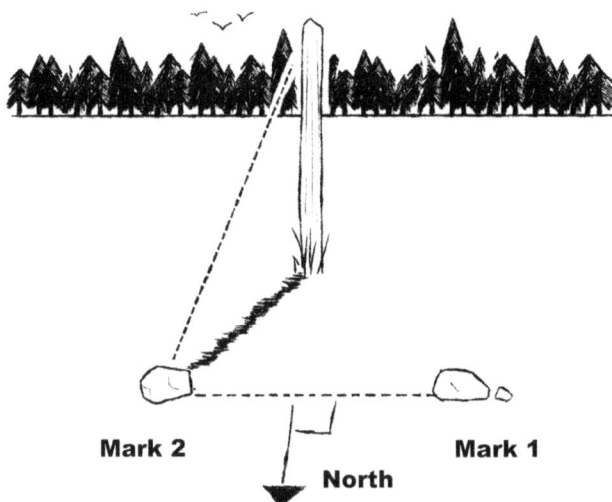

Mark 2   **Mark 1**

North

Clock Face Method One

To use this method, you'll need an analog watch.

First make sure your watch is not set to daylight savings or any other obscure timings.

Hold your wristwatch in front of you like a compass and place a twig at its edge, so the twig casts a shadow towards the watch's center.

Turn your watch until the shadow splits the distance between the hour hand and 12 on the watch face in half. In the Northern Hemisphere, 12 o'clock will now be pointing south and 6 o'clock will be pointing north. In the Southern Hemisphere, it will be the opposite.

Clock Face Method Two

For a faster but less accurate method, point your watch's hour mark toward the sun. The center of the angle between the hour hand and the 12 o'clock mark is the north-south line.

In the Southern Hemisphere, do the same, but point the 12 mark to the sun.

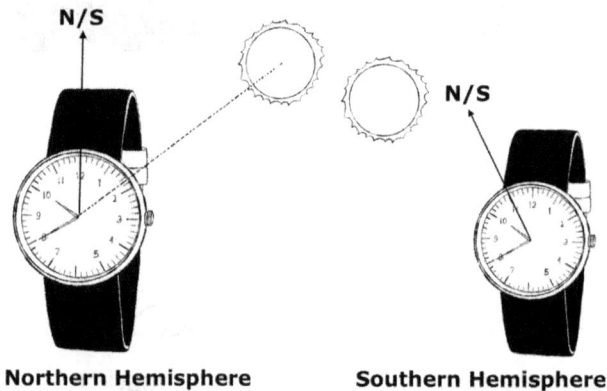

Northern Hemisphere   Southern Hemisphere

Clock Face Method Three

This method is for when you don't have an analog watch, but you know the time.

Draw a big circle on the ground, and then draw a straight line from the center of the circle straight towards the sun. When you're in the Northern Hemisphere, this is your hour hand.

Next, draw a line to 12 o'clock on the circle where it would be in relation to the hour hand. The point halfway between the two lines is roughly the north-south line.

When you're in the Southern Hemisphere, the line you draw to the sun is your 12 o'clock. Draw the second line as the hour hand. The point halfway between the two lines is roughly the north-south line.

Now you know the north-south line. Confirm north using basic sun movement (sun sets in the west, or any other method).

Vegetation

Using vegetation to tell direction is not as accurate as other methods, but it's good for confirmation or when you don't have time, such as when you're on the run.

Things get more sun on the side that faces the equator, so whenever you spot the following, chances are you're looking at the side faces the equator:

- Trees will have more branches, and the branches will be more horizontal (in comparison to the other side).
- On tree stumps, the rings will be spaced wider.
- Fruit will be riper.
- Vegetation will be thicker.

Wind

Here are several ways you can use the wind to give you general direction:

- Bends in trees are due to prevailing winds.
- Birds build their nests so they are protected from the wind.
- During the day, the breeze comes from the sea (or a large body of water) if there is one nearby. It is the opposite at night.
- Prevailing winds, such as a regular afternoon sea breeze, usually come from the same direction.
- Spiders build their webs so they are sheltered from the wind.

- The scent (sea, vegetation, or cooking, for example) carried by the wind can indicate the direction it's blowing from.

To determine wind direction:

- Watch the general direction of clouds.
- Throw some sand, leaves, or grass into the air.
- Watch the treetops.

Polaris (North Star)

At night in the Northern Hemisphere, Polaris is a good indication of north. It's not the brightest stay in the sky, but it is the only one that doesn't move.

To find Polaris, first find the Big Dipper and Cassiopeia, which are easy-to-identify constellations. Once you have them, follow the "ladle" of the Big Dipper up about five times its length, which is around halfway to Cassiopeia. Imagine a line from Polaris straight down to a landmark you can see. Use the landmark to guide you north.

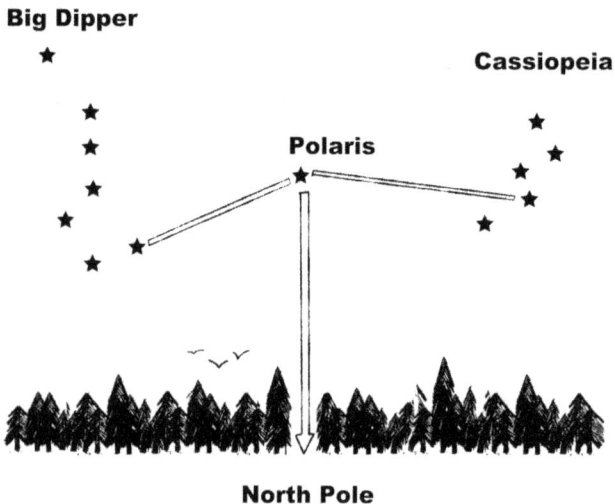

Southern Cross

The Southern Cross is the Southern Hemisphere's equivalent of Polaris, but it is a constellation of five stars as opposed to one. It is a fairly accurate way to indicate south.

The four brightest stars of the Southern Cross form a cross angled to one side. There are also two bright "pointer stars" to its southeast.

Do not confuse the Southern Cross with the False Cross cluster of stars. You can tell the difference because the False Cross has the following characteristics:

- It's bigger.
- It has more of a diamond shape.
- Its stars aren't as bright.
- There are only four stars.
- There are no pointer stars.

Once you find the Southern Cross, imagine a line in the direction of the cross points, and then another line at right angles between the two "pointer stars." Trace a third imaginary line from the intersection of these two lines straight down to a landmark you can see. Use the landmark to guide you south.

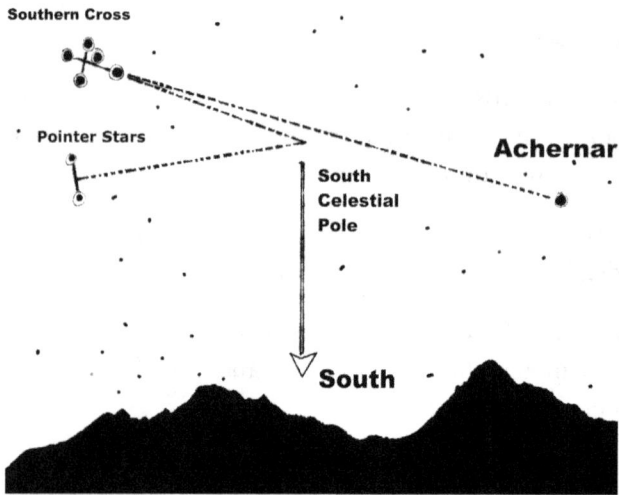

To confirm the intersection point is correct, you can draw an imaginary line five times the distance of the long axis of the cross, in the same direction. The bright star Achernar will be another five lengths after that.

Orion

The constellation of Orion is visible in both hemispheres. It has three brighter stars (Orion's belt) and several dimmer stars (his sword).

You can use Orion in the same way as basic sun movement, because it rises in the east and sets in the west. Orion changes orientation over the course of the night, starting horizontally.

East, Mid-Evening

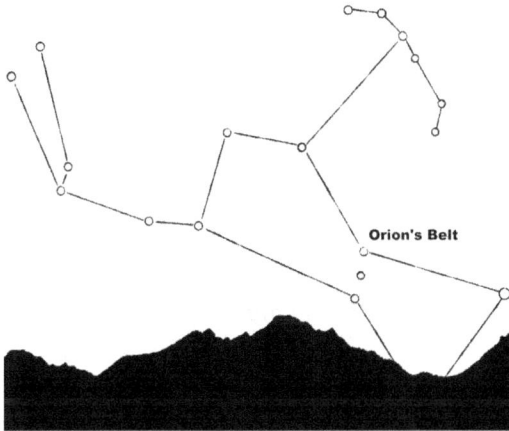

South, Late Night

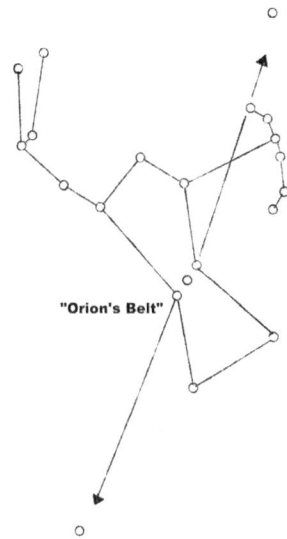

Orion's Belt

"Orion's Belt"

Navigation by Any Star

When you can't identify any of the above constellations, stick a tall stick (about 1m long) in the ground. Get another stick about half the size of the first one. Sight the tops of both sticks to a bright star, and stick the shorter stick in the ground. Come back in 30 minutes and notice in which direction the star has moved. In the Northern Hemisphere:

- Right = you're facing south.
- Left = you 're facing north.
- Up = you're facing east.
- Down = you're facing west.
- Right and up = you're facing southeast.
- Right and down = you're facing southwest.
- Left and up = you're facing northeast.
- Left and down = you're facing northwest.

The directions are opposite in the Southern Hemisphere.

Moon

There are two basic methods for determining direction with the moon. The first one relies on the fact that the illuminated side of the moon is always nearest the sun. Between sunset and midnight, this illuminated side will face west. Between midnight and sunrise, it will face east.

The second method is to imagine a line that joins the tips of a crescent moon and goes to the ground. This will give you south if you're in the Northern Hemisphere and north if you're in the Southern Hemisphere. The higher the moon is in the sky, the more accurate this will be.

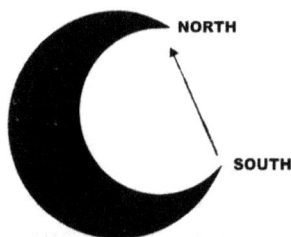

Improvised Compass

Any magnetized "needle" makes an improvised compass you can use to determine magnetic north. The needle can be any piece of iron, nickel, or steel that is long, thin, and light, such as a pin, needle, or straightened paperclip. The material must be able to rust. Aluminum and yellow metals (tin, copper, etc.) won't work.

There are a few ways to magnetize your needle:

Rubbing. Rub the needle on a magnet (extract one from a radio or pair of headphones), silk, or any synthetic fabric (nylon, Dacron, Kevlar, parachute material, etc.). Rub in the same direction 30+ times. At the end of each stroke, lift the needle 5cm (2in) in the air, and return to the beginning of the needle for the next stroke.

Electromagnetization. For this, you need a 2-volt (or higher) battery and some insulated wire. Coil the wire around a needle. If the wire is uninsulated, wrap the needle with paper or cardboard. Attach the ends of the wire to the battery terminals for five minutes.

Razor Blade. This method is not as effective as the other two, but is better than nothing. Carefully scrape the razor blade across the palm of your hand 30+ times.

Put your magnetized needle on top of a leaf (or something similar such as a woodchip, paper, etc.) and float it in still water. The water must not be in anything that can be magnetized, such as a tin can. Aluminum, plastic, a puddle on the ground, etc. are all good. Alternatively, suspend the needle on some string.

When the needle settles, it should point north/south. Disturb it a few times to see if it realigns in the same direction. Use other survival navigation methods to confirm which side points north and mark it. Top up the needle's magnetism several times a day.

Action Plan for When You're Lost

Stop as soon as you think you might be losing your way.

Look around for recognizable landmarks. If you can't find one immediately, try:

- Backtracking until you get to a known location.
- Seeking high ground for a better vantage point.
- Triangulation.

If you are completely lost, you need to make a plan. Consider:

- Whether you are being hunted or not.
- Making camp (time, weather, etc.).
- Rescue signals.

Travel in the direction settlements are most likely to be (downstream, for example).

To prevent yourself from getting lost:

- Periodically look back and take a mental note of what you see.
- Check your position regularly.
- Have navigation aids.
- Leave markers back to your campsite.
- Carry signaling devices, such as a whistle, flashlight, or radio.

MOVE SAFELY

In order to evade your enemy, you may have to go faster than usual, travel in the dark, or take other risks that you normally wouldn't.

However, if you get injured, it will slow you down and can leave additional clues your enemy can use to track you. In addition, a small injury in the wilderness can quickly turn into a major issue. For these reasons and more, it is important to know how to move safely through various terrains.

Jungle

Jungle is characterized by dense vegetation, heat, and humidity. There are plenty of plants and insects to sting or bite you, as well as some larger predators depending on your location.

To make your travel in the jungle safer:

- Be careful about touching any vegetation. Cover your skin with clothing as much as possible, and don't grab anything with your bare hands. Use a stick to move it out of your way instead.
- Check for and remove parasites, such as ticks or leeches, regularly.
- Despite not wanting to leave signs of your trail, you may need to chop through vegetation to get anywhere. If you do, cut it low and on a downward angle, so that it falls away from you as opposed to falling in your path.
- Following electric or telephone lines often makes for an an easier path, but be careful of running into your enemy if you do this.
- If you get snagged in thorns or other vegetation, stop and back up until you're free.
- Look through the foliage to find breaks in the jungle where you can move

- Look for and follow game trails to make movement easier, assuming they're going in the right direction.
- Move slowly, steadily, and smoothly. Don't force your way through. Find the easiest route and adapt to it.
- Only cut what is absolutely necessary, so you leave fewer signs of presence. Chopping unnecessarily will also wear you out faster.
- Only travel when it's light out. In most other terrains, night travel isn't too challenging, but don't attempt it in the jungle.
- Start setting up camp early, while you still have light. It gets dark under the canopy well before night falls.

Mountains

Traveling through mountainous terrain is often dangerous due to the risk of falling, either off the rocks or through ice. Another danger is the weather. It can turn without warning.

Higher ground is easier to navigate, but you're more visible to your enemy there, and food and water are less plentiful.

As a rule, avoid ice fields, loose rock, and scree.

Here are some safety tips for climbing slopes:

- Climb steep slopes in a zigzag manner.
- Drive your snow axe in sideways for stability.
- Never try to descend high cliffs, especially without a rope.
- On steep cliffs, face the rock.
- For less steep rock faces with deep ledges, adopt a sideways position and use the insides of your hands for support.
- When descending gentle slopes, face out, with your body bent. Dig your heels in and use a walking stick. If possible, carry your weight on the palms of your hands.
- When descending steep slopes, go backwards and drive a stick into the snow for support.

In mountainous terrain, you'll probably need to do some rock climbing. Follow these basic tips:

- Always have three points of contact.
- Do not climb higher than you're willing to fall.
- Keep your body away from the rock, use flat feet, and look up.
- Never pull outwards on a loose rock. If it falls, shout a warning to those below.
- Test each hold before committing to it.

When crossing glaciers:

- Probe the snow in front of you for crevices.
- Tie groups of people together at no less than 9m (30ft) intervals. Have a mainline which everyone ties onto, preferably one with a prusik hitch.

In case of an avalanche:

If it starts below your feet, get upslope of any cracks in the snow. When you are below it, move to the closest side out of its path.

If you can't avoid it, get rid of all excess weight and grab something solid, such as a tree. Do not abandon your ski pole or communication devices.

When there is nothing to grab onto, use the freestyle swimming stroke to stay on top of the snow. If you are unable to stay on top, put your hands in front of your nose and mouth to create an air pocket.

As soon as you stop, make as big an area as possible while trying to reach the surface. Use your ski pole to poke around and find open air. To figure out which way is up (to the surface), spit and go in the opposite direction from the one in which the spit falls.

Put your hands in front of your nose and mouth to create an air pocket.

Desert

Any large, dry area of land is a desert. It's characterized by little vegetation and extreme temperatures.

Never travel through it unless you are certain you have enough water to reach your destination. You also need something on wheels to haul your water and other supplies; otherwise, you'll lose more fluid than you can carry to replenish it.

To conserve water, travel at night and rest in the shade during the day. This also keeps your body temperature up during the cold nights. Ensure you have enough cold-weather clothing for night time.

The dessert will play tricks on you. Most people will underestimate distances. What you think is 1km will probably be 3km. Extreme heat will cause mirages. Combat them by surveying the area when the temperature is lower, such as at sunset.

It's easiest to move along valley floors between dunes.

Arctic and Subarctic

Arctic and subarctic terrains are cold deserts. The temperatures are extreme and travel through thick snow is tough.

To make traveling easier and safer:

- Always cross a snow bridge at right angles to the obstacle it crosses. Find the strongest part of the bridge by poking ahead of you with something, such as a pole or ice axe. Distribute your weight by crawling, or by wearing snowshoes or skis.
- Compasses are not accurate near the poles (north and south). Use survival navigation instead.
- Do not travel when visibility is poor or when there are very cold winds.
- Snow is firmer at dusk and dawn.
- Use a walking stick to probe for pitfalls.
- When you're on the water, avoid icebergs and sailing too close to ice cliffs.

A pair of snowshoes will make movement easier due to the increase in surface area. Improvise them from green saplings and paracord (or whatever you have).

Canadian emergency snowshoes are good to make when cordage is limited. To do that, you'll need:

- 10 x 1.5m (5ft) poles about as thick as your thumb.
- 10 x 25cm (10inch) sticks about as thick as your thumb.
- Cord.

To construct the shoes:

- Lay five of the poles next to each other so their bases are spaced evenly along one of the sticks.
- Tie them in place.

- Tie another stick about halfway up the poles, and a third stick a few centimeters from that one. Sticks two and three are where your heel will sit. Tie sticks four and five where your toes will be.
- Tie the tips of the pole together.
- Make the second shoe in the same way.

Tie your normal shoes to the improvised snowshoes.

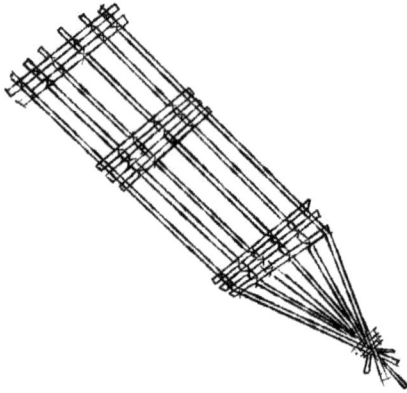

Here are a couple of other designs you can use, depending on your resources.

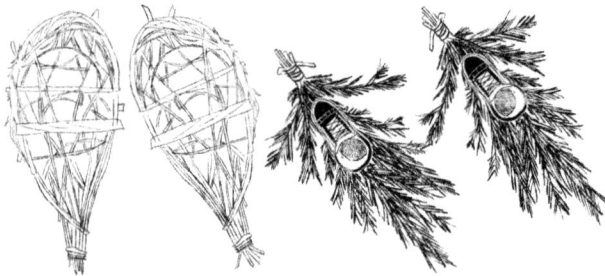

To walk in snow shoes, lift them high enough to clear the surface of the snow. Make sure your feet hold most your weight with each step. That is, don't balance on the tip and/or tail of the shoes.

When you have a lot of gear, you can make a travois, which is a sled/sledge for carrying stuff (as opposed to riding). To make one, you'll need sticks and cordage:

- Find two forked branches about the same size. The non-forked portions are the parts that will carry stuff.
- Find seven or more sticks of the width you want. The exact number of sticks will depend on how long the travois needs to be.
- Remove one side of the fork on both branches. These branches are the runners.
- Lay the runners parallel to each other at the width you want.
- Tie three of the sticks to the runners. Space them evenly, so they make squares.
- Tie two cross-braces on each square to create triangles.

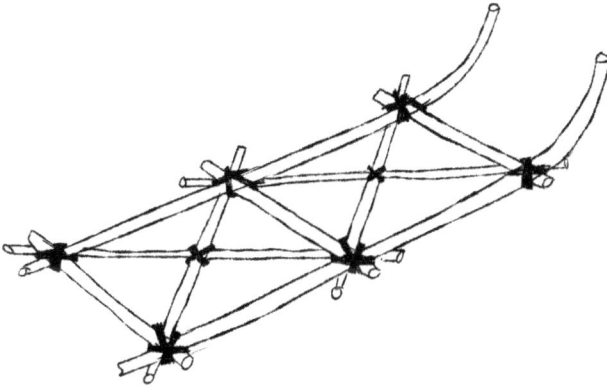

Ice

Crossing ice is a risk you should avoid. When you don't have a choice, test each step carefully before committing. If you're worried about the ice breaking, distribute your weight by lying flat and crawling.

To recognize thin ice:

- A thin area of ice covered in snow will be darker than the surrounding area. A thin area with no snow will be lighter.
- Contrasting colors in the snow/ice indicate thin ice.
- Ice is weaker near a river's mouth or a shore.
- New ice is generally thicker than old ice.
- River ice is generally weaker than lake ice.
- Snow-covered ice will be weaker. The snow insulates it from freezing and adds weight.

Escaping from a fall into ice water is not easy, and the result can be deadly.

DO NOT PRACTICE THIS IN ICE WATER! Go through the motions in a pool instead.

When you first fall into ice water, you'll start to hyperventilate. Try to stay calm and keep your head above the water. Taking deep breaths may help, but be careful not to breathe in any of the water.

After one to three minutes, the shock response will begin to wear off. Now you'll have about 10 minutes to get out before you fall unconscious. Once you have your hyperventilation under control, find where you first fell in. You want to get out where you know the ice was strong enough to support your weight, so going back to where you came from is your best bet.

Place your hands on the surface and pull yourself up, while staying as flat to the ice as possible. Pulling yourself straight up will be far less effective and a waste of energy.

Kick your legs as you creep out of the water. It will be very slippery.

Once you are out of the water, lie flat on the ice and roll away. Rolling keeps your weight distributed, and has less of a chance of creating further cracks in the ice.

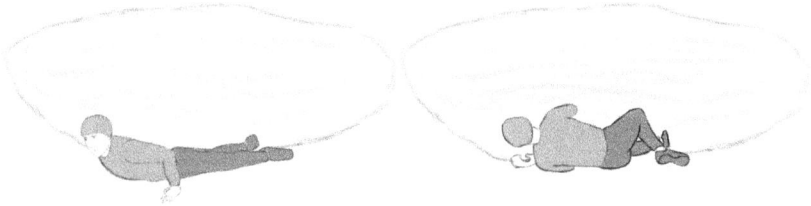

If you know you'll be crossing ice country, it's very wise to get some ice picks. They will make it far easier to pull yourself out of the water, although it will still be difficult.

At the very least, put some large nails in your pocket to help you grip the ice when you need to pull yourself out.

If you can't get out, you need to conserve your heat and energy. Put your arms on the ice and keep them there, so that they freeze to the surface. That way, when you lose consciousness, you'll have a better chance of not falling into the water.

Never go out to someone who has fallen into ice. Coach them on what to do from a safe distance, and hold something out for them to grab onto, such as a stick or a rope.

Once you're out of the water, get out of your wet clothes and get warm as soon as possible.

Cars and light trucks need at least eight inches of clear, solid ice to drive safely. If you're driving on ice, do so slowly, and don't stop until you're clear of it. Cross any cracks at right angles, and leave at least one vehicle length between vehicles.

Rivers and Streams

Following a river downstream is a good way to find people and slower water, except in the Arctic if the river is flowing north.

Avoid thick vegetation or other obstacles that make it hard to follow the river by seeking higher ground and then cutting the river off at the bends. As well, avoid dry creek bottoms and ravines with no escape, in case of a flash flood.

When crossing a stream or river, do so at the safest possible place. Unless you can jump it, narrow is not best. Look for straight, wide, and shallow water. The current is faster at the bends and usually deeper in narrow channels. Lots of debris is also a sign of fast flow. Test the current by throwing a branch in and seeing how fast it goes. Mild ripples are generally safe to cross. Whitecaps (small, surface-breaking waves) will be slippery.

Although it may be wider, the point where a river breaks into channels is usually a good crossing location. The energy of the current dissipates there, and there may be small patches of land where you can take breaks.

Check 100m (33ft) downstream of where you plan to cross. Make sure there aren't any hazards you could be swept into. Consider your entry and exit points. An easy exit point is especially important. You want something low and open so you don't have to climb through or up anything.

You may be able to avoid getting wet if you find a fallen log that spans the width of the river. If you do find one, don't try to walk over it. It is much safer to straddle it and scoot yourself across. You must be very sure it will hold your weight. When you're in a group, have only one person cross at a time.

Wading is a method of walking through water. Use it in water no deeper than thigh height. When wading, you can take your pants, shirt, and socks off to lessen the water's drag. Doing this will also give you dry clothes on the other side.

Keep your shoes on. You won't want to risk damaging your feet, and you will want the traction shoes will provide.

Tie your clothing to the top of your pack, or in one bundle if you don't have a pack. The idea is to keep everything together so it's easier to find if you have to discard it while crossing.

If you're wading across with a pack, carry it on your shoulders. Leave your waist belt unclipped and loosen your shoulder straps, so you can discard the pack if necessary. You won't want to be struggling to get it off if the current sweeps you off your feet. Don't worry about having a heavy pack when crossing. It will keep you more stable.

If you're wading alone, use a strong branch to support you as you cross. Three legs are far more stable than two.

Position yourself upstream of your chosen exit point so that you can cross at a 45-degree angle to the current. Face upstream and place the tip of your branch on the bottom of the river in front of you. Keep it slanted and let the current push it against your shoulder. Your branch will break the current and provide stability. Shuffle sideways and a little downstream across the river. Use small, low steps. Do not cross your feet.

Always maintain at least two points of contact with the bottom of the river. Don't move too far on either side of your branch. You don't want to be leaning.

Only reposition the branch once your feet are very stable on the river floor. Shift it in small increments, feeling for your next placement. Lift it off the river floor only as much as you need to.

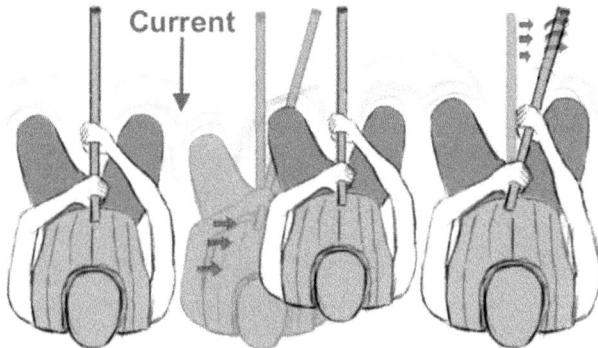

At Sea

When at sea in a raft, use either the current or the wind to take you to where you want to go.

To use the current, deflate your raft a little so it rides low in the water. Deploy your sea anchor and keep low in the raft.

To use the wind, you need a sail. Inflate the raft so it rides higher, pull the sea anchor in, and sit up so your body catches the wind.

If you improvise a sail, prevent the raft from capsizing by holding the bottom with your hands, so you can release it quickly if there is a sudden gust of wind.

When in rough waters, keep low and stream the sea anchor from the bow (front). Tying life vessels together will improve their stability.

Look for the following to indicate land:

- A constant wind with a decreasing swell. Land is windward.
- A green tint on the undersides of clouds.
- Isolated cumulus clouds.
- Muddy water, indicating silt from a large river mouth.
- Lighter-colored water, indicating shallow water.
- Seabirds flying. They move away from land before noon and return to it in the afternoon.
- Odors and sounds of land, including smoke, vegetation, surf, animals, etc.

Cumulus clouds are "puffy" with flat bases.

Once you find land, you need to get to it safely. Wait until daytime to choose a landing point, and select one from which it will be easy to beach or swim ashore. The downwind side of an island is usually better.

As you approach the shore, assess the features of the landscape (high ground, vegetation, water courses, etc.). Choose a meeting point in case you get separated.

Secure all your gear to your body and have a flotation aid ready. Stay in the raft for as long as possible.

Take down the sail and put a sea anchor out to keep you pointing at the shore, unless you're going through coral. Head for gaps in the surf. Waves usually occur in sets of seven, from small to large. Steer clear of rocks, ice, and other obstacles.

When you get close, paddle hard and use the waves to carry you into shore. If the surf is heavy, point towards the sea and paddle into approaching waves. To avoid getting swept back out to sea, make the raft as light as possible and take out the sea anchor.

When the undercurrent tries to take you back out, fill part of the raft with water and stream the anchor towards the shore.

Learn more about water safety and survival in *Survival Swimming*:

www.SFNonfictionBooks.com/Survival-Swimming

Shorelines

When moving along shorelines, be careful that the tide doesn't cut you off. You can see how far the high tide will go up the shore by:

- The line of debris (weeds, shells, trash, etc.).
- Changes in sand texture.
- Color changes on rocks/cliff faces.

Where a beach falls steeply into deep water, there will be a strong undertow. If you must go in, have a safety line anchoring you to the shore.

Group Travel

When traveling in a group, ensure everyone knows the route and rally points in case you get split up.

Have one or two people (scouts) go ahead to find the best routes. Appoint someone who is back with the main group to ensure the scout(s) maintain direction. Relieve the scouts frequently.

During rest times, wait for everyone to catch up. Check and adjust equipment, rehydrate and refuel, take care of injuries, etc. Remember, you are only as fast as your slowest members.

In covert situations, consider splitting into pairs. Two heads are better than one, but a large group is easier to track. Set a rally point if you want to meet up again later.

Related Chapters:

- Survival Navigation
- Clothing
- Insects

OVERCOME OBSTACLES

Obstacles are any things that will slow you down as you move, and/or places where you are more likely to be seen.

Avoid obstacles whenever possible, especially ones that are inherently dangerous in themselves. The only exception is night movement. It's better to move at night, except when the terrain doesn't allow it.

Observe an obstacle from a distance before crossing it. Look for best way to cross and best time to move.

When it comes to stealth, there is an order of preference of how to cross obstacles. The one you choose depends on the difficulty of doing it and the time factor.

- **Around**. If won't add risky exposure (e.g., light, time).
- **Under**. Dig, or lift the bottom of it.
- **Through**. Find a weak point and cut a hole if needed.
- **Over**. Cross quickly and keep your profile as low as possible. To prevent injury, land on two feet and roll if necessary.

Night

When moving at night, you need to compromise between the easiest and safest routes.

Avoid using light, especially white light. Memorize your route to minimize the need to refer to your map.

A half-moon provides a good amount of light for stealth movement. It lets you see where you're going while keeping you hidden.

Stairs

Move along the edges of stairs closest to the wall. The middle will make more noise.

Around Corners

Lie flat and look around the corner. Do not expose yourself any more than necessary.

Windows and Mirrors

Stay close to the side of the building and pass below the window/mirror level.

Wire Fences/Obstacles

Ensure fences are not electrified or fitted with other security devices. Look for:

- Warning signs.
- Bare wires going into insulators.
- Small, dead animals.

To go under a wire, slide headfirst on your back by pushing forward with your heels. Place a length of wood (or something similar) lengthwise on your body so the wire slides along it. Feel ahead with your free hand to find the next strand of wire, if there is one.

When going it under isn't practical, try going through it. Cut the lower strands so there are fewer sign of tampering. To do this quietly, hold the wire near its support and cut between your hand and the support. This technique also prevents the ends from flying away.

For even less noise, partially cut the wire and finish it off by bending it back and forth. If needed, stake the wire back to allow room to crawl through

If there's a low wire obstacle, step over it carefully. To climb over taller ones, find holds near the support posts.

In the case of barbed wire, you need to take extra care you don't get snagged. Before climbing over, cover the wire with any flat, heavy material, such as:

- Carpet.
- Thick blanket.
- Several layers of cardboard.

Razor wire is very dangerous. If you absolutely have no other choice, use a curved stick to pull wire down flat and cover it with heavy material before climbing over.

Solid Wall

If you can't go around, under, or through it, find a low spot to climb over.

Test the integrity of the wall by grasping it and lightly pulling it straight down. Gradually increase strength until you're lifting your body off the ground.

Check if other side is clear (if possible), and if it is, roll over the wall as quickly as possible.

To learn how to run up high walls and overcome other obstacles, check out *Essential Parkour Training*:

www.SFNonfictionBooks.com/Essential-Parkour-Training

Open Areas

Open areas are those that have little to no cover, like grass fields. Only cross them if there is no other practical way around.

To cross open areas, choose the lowest ground possible (furrows, for example) and lower your profile as much as is practical. Consider speed vs the need for concealment.

In grass, try to time your movement to when the wind is blowing, and change direction slightly from time to time as you cross. This helps to cover up the path of your movement.

Roads, Trails, and Railroad Tracks

Never move along roads in a covert situation. To cross them, use narrow points with low traffic and concealment to minimize your exposure (bushes, shadows, a bend in the road, low ground, etc.).

Use a low run to cross them.

Be careful of areas with no traffic, as they may be booby-trapped.

Caution: If there are three rails on the railway tracks, one may be electrified.

In Public but Hostile Territory

Avoid contact with the locals, especially children and dogs. Go around populous areas if possible.

Do your best to blend in before entering. Wear local clothing, cover your skin, get clean, etc.

Unless you are fluent in the local language, do not talk. Instead, look down and keep walking past anyone who tries to engage you.

Bridges

Avoid crossing bridges. It is better to swim across. You can hide underwater and use a reed or straw to breathe.

When the body of water is too dangerous, wait for an opportune time and cross the bridge as quickly as possible.

If you are caught on the bridge and death is imminent, jump into the water. This is very dangerous, especially if you do not know the depth or the water.

When jumping, try to land in the channel where boats go under the bridge. This area is generally in the center, away from the shoreline.

Stay away from any area with pylons that are supporting the bridge. Debris can collect in these areas, and you may hit it when you enter the water.

Jump in feet-first, keeping your body completely vertical. Squeeze your feet together, clench your backside, and protect your crotch with your hands.

After you enter the water, spread your arms and legs wide, and move them back and forth to slow your descent.

Related Chapters:

- Observation
- Clothing

PREDICT BAD WEATHER

Bad weather makes it harder for your enemy to track you, but very bad weather can be hazardous to your survival. Learning how to read nature's signs of bad weather will allow you to plan and prepare appropriately.

Individually, none of these signs are a particularly accurate way of predicting the weather. Use a few of them together for best results.

Clouds

There are many types of clouds, but it's best to keep things simple. Here are a few that will tell you the most about the oncoming weather.

In general, the higher the clouds, the finer the weather, while clouds that are low and dark usually bring rain. Consider the direction of wind to predict what weather is heading your way.

Cumulonimbus clouds are storm clouds. They're low, dark, and have flat tops, like anvils.

Cumulus clouds are white and fluffy. When separated, they bring fine weather. If they're large and clumped together, expect sudden showers.

A blanket of gray clouds brings drizzle, light to moderate rain, or snow.

Gray clouds in the evening bring rain, but in the morning, they indicate a dry day.

A cloudless sky at dusk means you're in for a cold night. If there's a clear sky at dawn, the day will probably be hot, unless it is the first day of a cold wave.

Wind

A sudden change in wind usually comes with a change of weather of some sort.

Winds from a specific direction often bring similar weather every time. For example, in the Northern Hemisphere, winds from the south often bring rain.

Fog and mist together create condensation, but mean that rain is unlikely. If a wind blows the fog away, there is a chance of rain.

Animals

Animals are still in tune with nature, and sense bad weather before most humans do. When rain is coming:

- Animals in general become noisier.
- Birds fly lower.

- Frogs stay in water.
- Herding animals gather and feed more.
- Insect activity increases (except among bees, which disappear).
- Seagulls stay close to land.
- Spiders stay in the centers of their webs.

Miscellaneous Signs of Bad Weather

Other signs of bad weather approaching include:

- Body aches and pains appearing.
- Camp fire smoke swirling and/or falling towards the ground.
- Contrail lines from jet aircraft not dissipating within two hours.
- The air becoming damp (the walls might "sweat," for example).
- Flower and tree leaves closing.
- Vision and hearing improving.
- Natural springs flowing faster.
- Rainbows appearing in the morning.
- There being a red sky in the morning (a red sky at night is usually followed by several clear days).
- Ropes swelling.
- Coronas (the circles that appear around the sun and the moon) shrinking.
- Sound carrying farther.
- The smell of vegetation becoming more distinctive.
- The temperature not dropping at night.

Noticing one or two of the signs above does not guarantee bad weather, but if you experience several of them, you can be pretty confident it's coming.

Signs of Weather Clearing

Here are some signs that the weather will clear soon.

- Bees reappearing.
- Clouds rising, breaking, or lightening in color.
- Patches of blue sky appearing through the rifts in the clouds.
- Raindrops growing smaller after a change of wind.
- Wind direction shifting.
- Snow fall getting thinner.
- Temperature falling rapidly.

SHELTERS

Building a shelter is not ideal in an evasive survival situation. It's time-consuming and can leave large signs of presence. But if you're on the run for long enough, you'll need to rest eventually, either to prevent injury or to protect yourself from extreme weather.

When the weather is fine, crawling into the thickest vegetation you can find will keep you hidden. If the vegetation is thick enough, it will even keep light rain off you. Another option is to dig a depression, crawl in, and cover yourself with foliage.

When the weather is extreme, you need to construct a shelter. The key to a good evasive shelter is to keep it simple. You need it to be fast to make and dismantle, and to show the least sign of presence while you occupy it and after you leave.

Specific ways to make shelters are detailed in the following chapters, but there are some extra points you need to know for desert and snow shelters first.

Desert Shelters

In the desert, it's vital to keep cool during the day. If you have some material to build your shelter, use it in a way that maximizes airflow:

- Place the material about 50cm (20in) from your head.
- Create a 40cm (15in) airspace between the sheets. Avoid cutting the material. Fold it in half instead.
- Arrange it so the lightest side faces out to reflect heat, but only do so if it won't attract your enemy.

Snow Shelters

In the snow, you need to stay warm when you're not moving, and you may need to make snow blocks to construct your shelter.

Choose snow that you can cut, but that's also strong enough to support your weight. Make the blocks 50cm x 50cm (20in) and 15cm (6in) thick. When constructing your snow shelter, ensure the entrance faces away from the wind. Insulate the floors with vegetation (or whatever is available) and pile snow around the sides.

Brush off all snow and frost before entering, and keep a shovel close by in case you need to dig yourself out.

It's best to keep your shelter at least 5m (15ft) from the edge of a body of water, even if that water is frozen. The freezing and thawing of ice/water will change its level.

CLOTHING

Adequate clothing can prevent the need to make a shelter, and can help to protect you when you're on the move. Here are some general tips to get the most out of clothing, as well as how to improvise it.

No matter what environment you're in, loose-fitting clothing is best. Cover as much of your skin as possible and take care of what you have. Keeping your clothes clean and dry will prolong their life. Make any repairs as soon as possible to prevent the damage getting worse, and always keep your clothes off the ground and shake them well before putting them on.

Dressing for Warmth

To maximize insulation, use the layer system. The more layers you have, the warmer you will be. If that is not enough, stuff dry insulating materials (leaves, grass, feathers, moss, paper, car seat foam, etc.) between your layers of clothing.

Outer garments should be windproof, but not waterproof. Animal skins are ideal and wool is better than cotton. If clothes made from improved materials are available, use them.

Waterproofing

Plastic is a good waterproofing material, but it doesn't breathe. Use it to protect you from the rain, but be careful of it hindering ventilation.

When plastic isn't available, use large sections of birch bark. Discard the outer bark and insert the inner layer under your outer clothing.

Gaiters

Gaiters protect your lower legs from insects, low-lying foliage, sand, snow, etc. Most materials will work. Wrap them around your legs and tie them in place.

Headwear

Wearing a hat protects you from the sun, and will prevent body heat escaping through your head.

You can improvise headwear with some cord, a handkerchief, and a 120cm x 120cm (50in x 50in) piece of cloth. This design protects you from the sun while ensuring ventilation, which is important in the desert and other hot areas.

- Make the handkerchief into a wad on top of your head.
- Fold the cloth diagonally into a triangle and place it over the handkerchief, with the long edge forward.
- Secure it around your head with a piece of cord.

If you only have one piece of material, you can forfeit the ventilation gap and use it as a keffiyeh. Fold the material diagonally into a triangle and place it over your head, with the long edge forward. Fold the left side of the cloth over to the right side of your face, just below your eyes.

Wrap it around your head and tuck it in. Do the same thing with the right side going left.

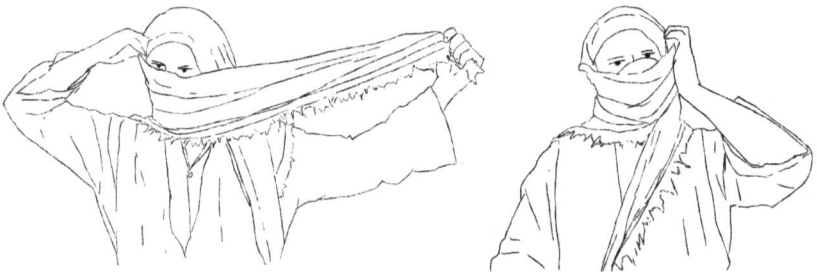

You can have your mouth and/or nose covered or pull that part down.

Mosquito Protection

Mosquitoes are annoying and a health risk. The best protection from them is to keep covered, especially at dusk, dawn, and during the night. When your clothing is inadequate, covering exposed skin with oil, fat, or mud may help.

Campfire smoke deters mosquitoes and other bugs, but is not a good idea in a covert situation.

Poncho

Improvise a poncho with any piece of material that's large enough, such as a blanket or bedsheet. Use plastic if you want it to be water-proof. Find the center of the material and cut a hole for your head to go through.

Shoes

Protecting your feet is important. Make improvised moccasins with two pieces of fabric 1m squared (more layers are better) and cord (optional).

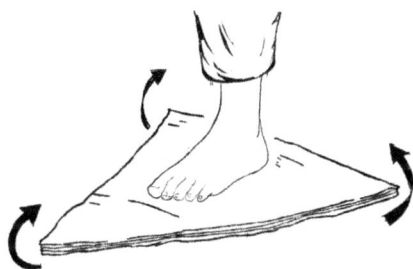

- If you have multiple layers, place them on top of each other.
- Fold the layers together into a triangle.
- Place your foot in the center, with your toes facing the corner.

- Fold the front over your toes, then fold the side corners over your instep.
- Secure each shoe with cord or by tucking the layers into each other.

You can make thicker soles from other material, such as rubber tires or bark. When you have soles and cord but no fabric, make holes around the edges of the soles and tie them on like flip-flops.

Skirt

Skirts are good for warm climates. Wrap any piece of material large enough around yourself like a sarong. Tie it in place if you have cord. Alternatively, tear leaves and fibers into long strips and tie them around a cord "belt" so they hang down like a hula skirt.

Sun/Snow Glasses

Glasses help to protect your eyes from dust, glare, and other things. To make improvised shades:

- Find a strip of fabric wide enough to cover your eyes and long enough to tie around your head.
- Place the middle of the fabric between your eyes and mark where your eyes sit.
- Cut small, horizontal slits where you marked it.
- Tie the fabric around your head so you can see through the slits.

When you don't have any spare fabric, use bark. Another way to reduce glare is to put soot under your eyes.

Related Chapters:

- Cord

SHELTER SECURITY

You're at your most vulnerable when you are asleep, especially if you're being hunted. Hiding your shelter is your first line of defense. Camp in places that are unlikely to be searched, and do everything you can to minimize the signs of your presence:

- Build it low and small.
- Camouflage the roof and sides with vegetation from the surrounding area.
- Cover your entry tracks (brush them away with a leafy branch, for example).
- Gather food and water, cook, clean, etc. away from where you're going to sleep.

You also need to consider natural hazards, such as falling deadwood, riverbeds, avalanches, trails (animal or human), and other things depending on your location.

A well-hidden shelter is good, but a little extra security goes a long way. Construct your shelter so you can see your enemy's likely direction of approach, and keep your stuff packed in case you need to leave in a hurry. Plan multiple escape routes, and sleep with a weapon.

When you're in a group, determine your rally points and have at least one lookout at all times.

Dismantle your shelter when you leave, so it looks like you were never there. Mask your scent by covering the area you slept on with soil and debris.

Button Hook

A button hook is a type of deceptive track. It allows you to observe your path of approach so you can see your tracker before he gets to you. If you see your enemy, you can either escape or ambush him. To do a button hook, walk straight past your shelter and circle back in a J.

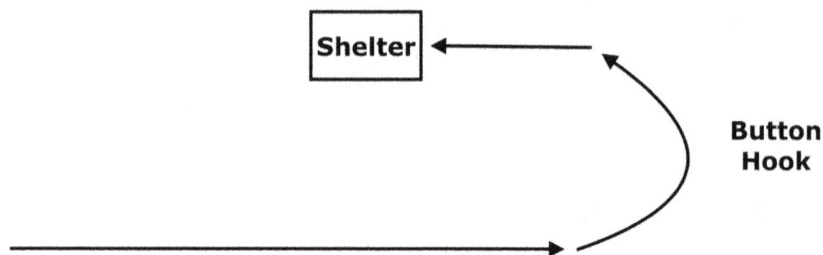

Path Guards

Path guards are warning systems or traps that protect you from approaching predators, whether human or animal. Set them up on all likely paths of approach in a way that uses natural obstacles to funnel predators into them, as explained below. Be careful you don't run into your own path guards. If you're in a group, ensure everyone knows exactly where the guards are.

Dismantle and cover up any signs of path guards before you leave.

Here are some simple path guards. More complicated ones (e.g., tripwires) are great for long-term security, but are not worth the effort for the quick "sleep and go" purposes of the evasive survivor.

Floor cruncher. Line the floor with something that will crunch when stepped on.

Red flag. This warns you if someone was there while you were gone. It is anything you place that has been moved. The item must be something that an intruder would think would normally be there, such as a branch lying across a trail.

Pitfall trap. Use the pitfall trap if you want to injure your enemy. Dig a small hole and stick wood spikes in it. Cover the top of it with foliage from the surrounding area so it blends in. You can coat the tips of the spikes with natural poison from plants or frogs if you have the knowledge.

Related Chapters:

- Camouflage
- Evade Trackers

EXISTING SHELTERS

Finding an existing shelter will save you time and energy, but it can be an obvious place for your enemy to look. Weigh the cost and benefit carefully.

Ready-Made Shelter

This is any shelter you don't need to do anything to, such as an abandoned vehicle, a spot under bridges, or an abandoned hut.

Be careful of existing inhabitants. Look carefully for animal or human tracks and nearby feces. If it looks like predator, such as a bear, sleeps there, find somewhere else.

Cave Shelter

A cave is a type of ready-made shelter, but one with a few extra things to look out for:

- Being cut off by the tide.
- Flooding.
- Rockfall.

If you have a choice, pick a cave above a valley, so you'll stay dry when it rains.

For extra warmth, you can:

- Make walls across any entry points with rocks and/or logs.
- Insulate the ground.
- Create a fire. Do it at the back of the cave to retain heat. It will also drive out animals, so ensure they have an escape route.

Have an easily removable way to block the entrance, but be careful not to block ventilation.

Windbreak Shelter

A windbreak shelter is enough to protect you from windchill, but probably won't protect you from heavy rain. Potential windbreak shelters include fallen trees, boulders, hollow logs, the insides of bushes, rocky outcrops, dirt mounds, undercut banks, snow blocks, etc.

To increase your comfort, insulate the floor and dig a water runoff channel.

In a cold climate, place the entrance downwind to keep the chill out. In a warmer/tropical climate, place the entrance upwind to keep mosquitoes out.

Here is an example of a windbreak shelter from a fallen tree.

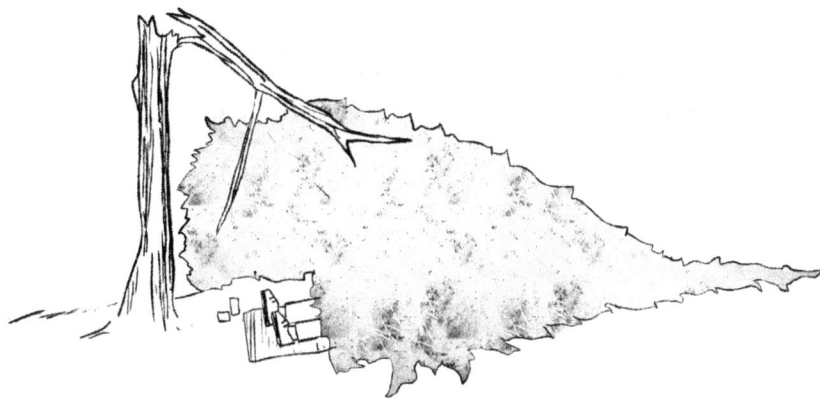

Tree-Pit Snow Shelter

This is an easy-to-build snow shelter that is hidden and gives you a 360-degree view. To make one, dig out the snow around a large tree. Use foliage to cover the top and to insulate the bottom.

To save work you can dig out one side only. Dig out the downwind side.

Open Ground

When on open ground, dig a depression and cover yourself with earth and foliage. At the very least, sit with your back to the wind.

LEAN-TO SHELTERS

A lean-to is basically an enhanced windbreak shelter, but you can also make it free-standing if necessary.

To make a lean-to, use a windbreak shelter as a base. Lay sticks up against it, then add overlapping green branches with the tips facing down, starting from the bottom. This will help shed rain.

Pile leaves around it for extra protection. Any large piece of material, such as a poncho, tarp, or survival blanket, can replace the foliage. Anchor it down with soil, rocks, or branches.

Materials that are not waterproof can still keep light rain out. Angle them steeply, double it up, and leave a small gap between the layers.

Camouflage the shelter as best you can.

Here are some examples of lean-to shelters.

Against a tree

Against a log

Self-standing

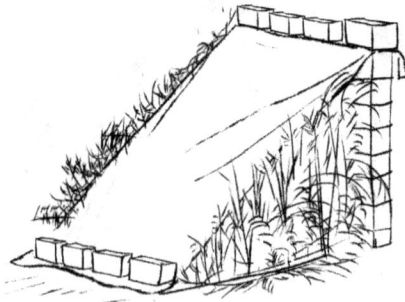

Using snow blocks

Related Chapters:

- Camouflage

TRENCH SHELTER

Trench shelters are good in places where there is little vegetation, like sandy environments. They work well in an evasive situation because they are hard to spot from afar, even if there is little cover.

To make a trench shelter, find a depression between dunes or rocks, or dig a trench long and wide enough for you to sleep in. Make it shallow, so it's less work to dig and harder to spot. Use what you dig out to build sides. Keep one short side open as an entrance. Make a roof out of sticks and vegetation or material, and camouflage the top and sides using the same material as the ground around it.

Here's one improvised for hot climates. It has a gap in the material to maximize cooling. Make it run north to south so it catches less sun.

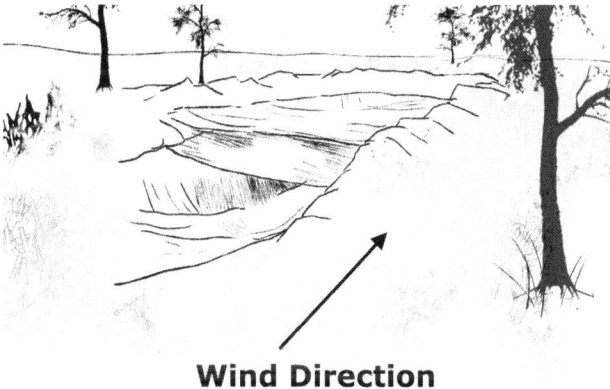

Wind Direction

In cold climates, orient the shelter so the wind hits the long sides and make the entrance at the lower end. In the arctic, create snow bricks and lean them against each other to form a roof. Fill any gaps with snow, and insulate the ground.

Wind Direction

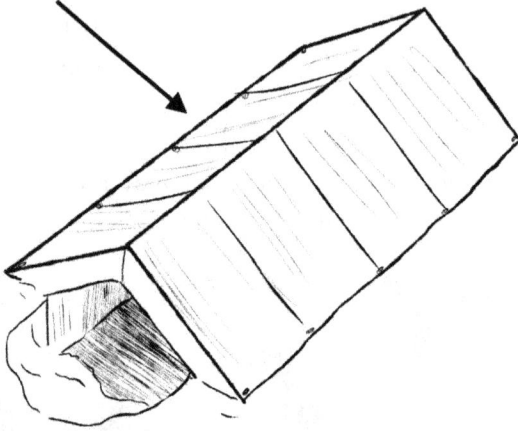

Related Chapters:

- Camouflage

IMPROVISED BEDS

In jungles, swamps, and similar terrains, it's better to sleep off the ground. This will keep you cool, dry, and undisturbed by crawling bugs.

Stacked Bed

To build a stacked bed, find four trees (or stick poles in the ground) in a rectangle, spaced widely enough to accommodate your body. Stack sticks and branches lengthwise on the insides of the trees until there's enough material to raise the sleeping surface as high as you need it (above the water level in a swamp, for instance).

A-frame Bed and Shelter

An A-frame bed is fairly easy to make and can double as a shelter.

First, create an A-frame by lashing poles together. Put two poles together side by side, so they lie horizontally. Tie the poles to the left of where you intend to make the rest of the lashing. Lay the short end horizontally between the two poles to the right of your knot, so you'll lash over them.

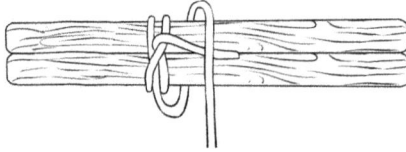

Wrap the end of the cord around the two poles. You need it tight, but not too tight.

Do at least as many turns as needed to make the lashing the same length as the width of the two poles.

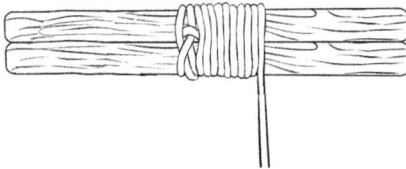

Do frapping turns by passing the cord between the two poles on the right side and then coming back up between them on the left.

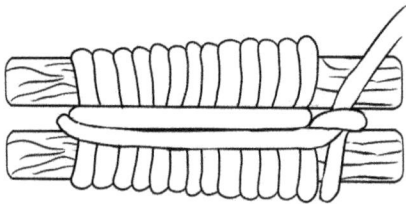

Do two frapping turns, and finish by tying the cord on one end around one of the poles. Pull the legs apart to make the A-frame.

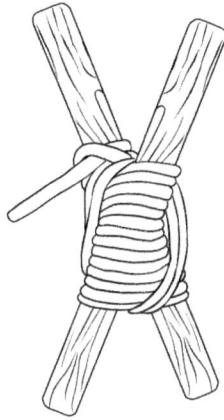

Lash two additional poles between the A-frames for the bed plat-form. Join these platforms with long poles to make your bed.

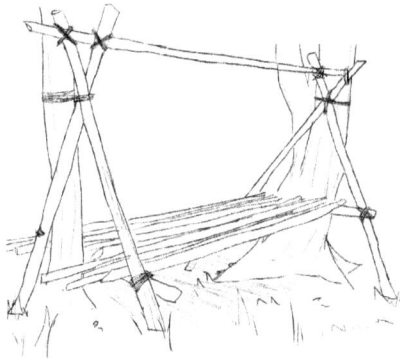

Make a cover for shelter and pad your bed (optional).

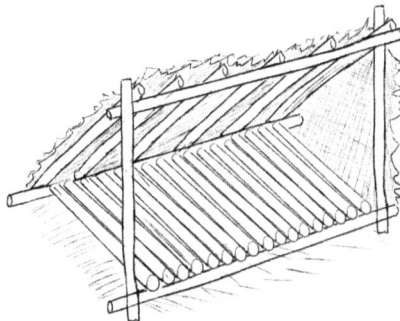

Tube Bed

A tube bed is faster to make and more comfortable, but you need a large piece of material to make it.

Find two poles long enough to be the length of your bed. Lay your material down and place the poles on it, spaced widely enough for you to sleep on. Wrap the material around them.

Build or find a couple of "shelves" to place your bed on so it's suspended off the ground. When you sleep on it, your weight will hold the material in place.

This variation of the tube bed uses the A-frame to hold the material out and keep the bed from slipping down.

Triangle Bed

The triangle bed is a way to suspend a bed off the ground without making A-frames.

Find three trees in a triangle. At least two of them need to be approximately the same height as your body. Lash a triangular frame between the trees at least 1m (3ft) from the ground (or water if in a swamp). Create your platform with material or foliage.

Hammocks

The common hammock is easy to make if you have a large piece of material and cord.

Find two trees far enough apart to fit your body height. Place two small stones in opposite corners of your material. Fold the corners over the stones, and tie cord around the material. The stones act as a stopper to prevent the cord from slipping off.

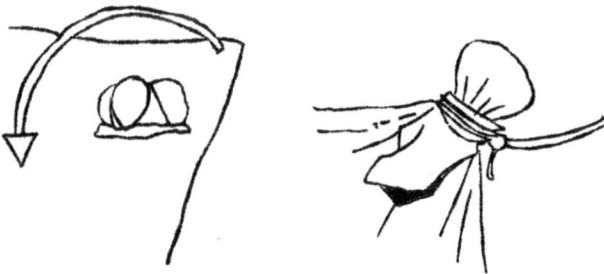

Tie the other end of the cord around the trees.

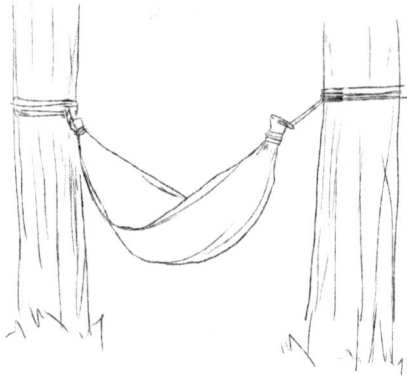

Here is a variation for when you have three trees. It's essentially a triangle bed made from material.

Bamboo Hammock

When you don't have any material, you can make a hammock out of bamboo.

Cut a thick piece of live, green bamboo about 1m (3ft) longer than your height. Carve a section out of it, leaving 50cm (20in) at each end intact.

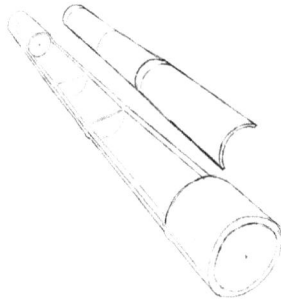

Tie cord around the ends to hang the hammock. This also reinforces the bamboo when you cut it. Make slits down the length of the bamboo, about 4cm (1.5in) apart.

Hang the hammock, then open it up and weave shorter lengths of bamboo between the slits.

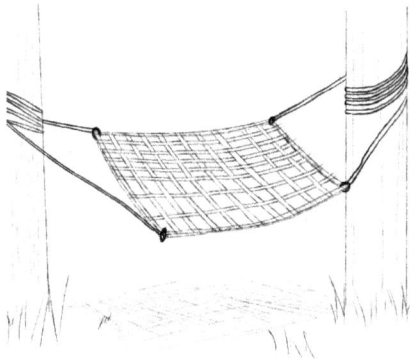

Add vegetation for comfort and insulation.

Fire Bed

A fire bed will keep you warm when you're sleeping on the ground, but it is not advisable to make one when you're trying to hide from your enemy.

Dig a 25cm (10in)-deep trench about the length and width of your torso.

Place a layer of dry fist-sized stones on the bottom, then make a fire on top of the rocks. Don't heat up wet, porous, slate, or softer rocks, or they may explode. If you aren't worried about enemy hearing you, bang them together before you use them to test if they're hollow or if they crack.

Once the fire has died down, spread the coals out over the rocks. Fill the trench up with soil and stomp it down. If coals come through or it is too hot, add more dirt. You do not want to catch fire while asleep!

Sleep with your torso over the heated ground.

Related Chapters:

- Cord

WATER

Water is essential for life and you will not last long without it, especially while running from your enemy.

Although it's not always easy, there are ways to acquire water in any climate. Unfortunately, many sources of water are not suitable for drinking, and using them can make you ill. Therefore, you must learn how to find water and how to treat it.

There are two basic ways to treat water: filtration and purification. A good water filter can eliminate many harmful types of bacteria, but it will not eliminate viruses. Purification will kill the viruses.

Ideally (besides having fresh drinking water), you'll filter water first and then purify it. When doing both is not possible, one or the other is better than nothing. Always treat the clearest water you have available.

CONSERVING WATER

When water is abundant, drink a minimum of one liter a day. When you're active, you'll need more. This is true no matter what the climate. Your body still loses liquid when it's cold, and dehydration can kill.

In all other cases, ration the water until you find a source of replenishment.

When water is scarce, (in the desert or at sea, for example) do the following to conserve what your body has:

- Cool your body with breezes and non-drinking water.
- Don't eat when nauseated. If you throw up, you'll lose water and any food.
- Don't smoke or drink any diuretics like alcohol or coffee.
- Eat less. Food requires water for digestion.
- Keep your body well shaded from above and below, if applicable. Avoid exposing yourself to reflections off water, for example. Cover as much of your body as you can.
- Keep your mouth closed. Don't talk, and breathe through your nose.
- Rest in the shade during the day and move at night.
- Separate yourself from the hot ground by sitting 30cm (10in) above it—on a branch, for instance.
- Sip whatever water you do have slowly and frequently. Moisten your lips, tongue, and throat before swallowing.

When at sea, soak your clothes in the water, then wring them out and put them on again. Only do this occasionally; otherwise, you might get saltwater boils. Be careful not to get the bottom of your raft wet.

FINDING WATER

When you're on the run, finding fresh drinking water is ideal since purifying it takes time. But if you have the resources, you should treat it anyway. Do not drink alcohol, blood, urine, seawater, or any water with signs of death, such as a dead animal or a lack of vegetation, near it.

Rainwater

Rainwater is safe to drink except in special circumstances, such as a nuclear or bio-contamination situation. Collect it in any containers you have, or improvise containers. You can use:

- Bamboo sections.
- Bark.
- Plants with big leaves.

To collect rainwater that has already fallen, use the soak-and-squeeze method. Use any non-toxic material you have (fabric, dry grass, etc.) to soak up the rainwater, and squeeze it into your container. This is also good for gathering water from small crevices, such as holes in rocks. Similarly, you can use cord to absorb and direct running water (at the base of a cliff, for example) into your container.

When at sea, wash off your collection material with seawater before using it to collect rainwater, unless it's clean to begin with. A small amount of salt is negligible, but if the material is encrusted with dried salt, it will be a problem.

Morning Dew

Morning dew is safe to drink. Use the soak-and-squeeze method to collect it. Tie material around your ankles and walk through the grass to soak it up.

When at sea, create a sun shade at night and turn the edges of it up to collect dew. Dew will also form on the sides of the raft. Wipe it with material to get it.

Plants

Depending on where in the world you are, certain non-toxic plants can provide you with drinking water. The best time to collect it at first light.

Don't drink any milky sap unless you can confirm it's from a source, such as a barrel cactus, that's safe to drink from. Here are some examples of general and plant-specific collection methods. For more ideas, research the area you will be in to see what water-bearing plants grow there.

Condensation

This method takes 12 to 24 hours to produce results.

Wrap and tie clear plastic over the green leaves of non-toxic vegetation. The condensation that forms will taste like the plant, but you can drink it.

Aboveground Solar Still

This method also uses condensation.

Fill half a clear plastic bag with green, leafy, non-toxic vegetation. Make sure there are no sticks or sharp spines. Put a small, clean rock in it as well.

"Scoop" air into the bag (or face it towards the breeze) and secure it, with the intention of keeping the maximum amount of air inside it.

Place the bag in the sunlight on a slope, with the rock in the bottom corner and the opening of the bag in the corner above the rock. The water will collect around the rock.

Plastic-Bottle Solar Still

This is a condensation method you can use if you find a plastic bottle with a lid.

Cut the bottom off a clean plastic bottle, such as a 2L soda bottle, and fold the bottom lip back into the bottle. Ensure the lid is sealed tightly.

Place the bottle in the sun on any moist surface, such as grass. Water will collect in the lip. Drink it straight out of the bottle, being careful not to spill it.

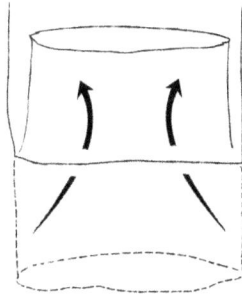

Tapping Trees

Many trees have sap, which is made up of water, minerals, and sugars. In most cases, it's safe to drink. Avoid evergreens (trees that have leaves all year round) and any trees that produce milky or dark liquid.

To tap the sap, hammer a knife into a large tree (preferably above a large root). Go about 5cm (2in) deep, on an upward angle. Take the knife out, and sap will drip from the cut.

Direct the sap into your container by placing something into the bottom of the cut, like sticks or grass.

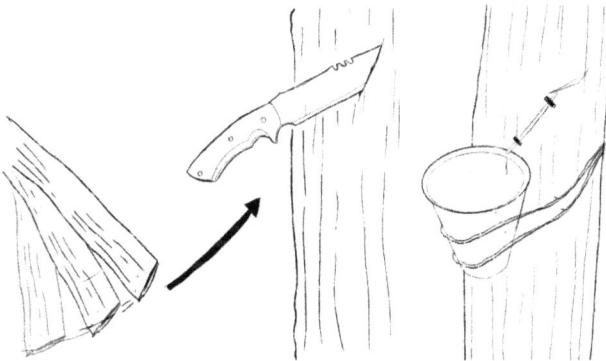

Drink the sap within 24 hours, before it starts to ferment.

Pulpy Plants

Many plants with moist, pulpy centers can provide water. Cut off a section of the plant and squeeze or smash the pulp so the moisture runs out.

Bamboo

Mature bamboo may hold water in its hollow sections. Shake it and listen.

When you find it a section that has water, cut a hole at the base of it and catch the water as is flows out. Alternatively, cut it at the top of a section and use a smaller piece of bamboo as a straw to drink it.

You can cut entire sections out and carry them for later.

Green bamboo thickets are good for collecting water if you have the time to wait. Bend a green bamboo stalk and tie it down. Cut off the top and let the water drip into a container.

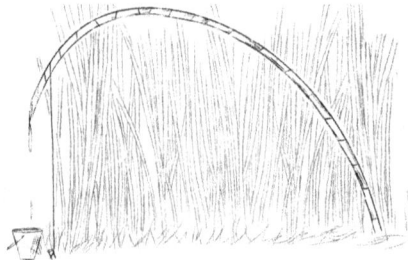

Nipa Palms

Nipa palms have a sugary fluid that you can drink. You can collect up to one liter a day.

Bend a flowering stalk down and cut off its tip. Cut a thin slice off the stalk every 12 hours to keep it flowing.

You can also eat the immature, jelly-like fruits, make teas from the flower petals, and cook the seeds of mature fruits.

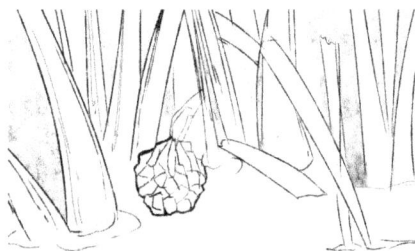

Other Palms

Other palms (buri, sugar, etc.) also contain a sugary fluid.

Press a frond lower to the ground with a blunt object. Bend the frond down, and the fluid will leak out where you hit it.

Coconuts

The liquid in green (unripe) coconuts is drinkable, but the liquid in mature coconuts can cause diarrhea.

Vines

Vines with rough bark and shoots about 5cm (2in) thick are likely to hold liquid.

Cut the bark off the vine to ensure the liquid is clear and water-like. If it is, make a notch in the vine as high as you can reach. Next, cut the vine off close to the ground and catch the water.

Do not:

- Drink the liquid if it is sticky, milky, or bitter.
- Touch the vine to your lips.

Banana/Plantain Trees

Banana and plantain trees can provide water for several days.

Cut down the tree, leaving a 30cm (1ft) stump. Scoop out the center of the stump so the hollow is bowl-shaped. It will fill with water.

At first, the water will be bitter. Discard it until it becomes palatable.

If you plan to come back to it later, cover the stump to prevent evaporation and contamination.

Plant Roots

Most plant roots contain water, but it's not always drinkable without treatment.

Dig up the roots and cut them into small pieces. Smash the pulp and catch the liquid.

Bodies of Water

Go to high ground to scan the area for bodies of water such as lakes, rivers, streams, and swamps. Be careful near them, as your enemy may be keeping an eye on them too. Animal predators also do the same thing.

Clear, cold, fast-running water from streams or waterfalls is the cleanest. Hold your container just below the surface, with the opening facing downstream. Still or slow-moving water must be treated.

Digging

During the dry season, try digging where there would be water, such as in dry river beds. Look for and dig in wet sand. When you can't find wet sand, try the outside edges of a sharp bend in a bed. Dig down at least 1.5m (5ft) to find seeping water.

Other places to dig are:

- Low ground, such as a valley.
- On the beach. When your beach well (or any other well) runs dry, dig deeper.
- Where there is green vegetation.

Animals

The behavior of animals can give you a good indication of where to find water. Here are some signs to look for:

- A column of ants may head to a small water source.
- Birds that are flying straight and low are usually headed towards water, especially in the early morning or late afternoon. If they frequently rest from tree to tree, they are usually heading away from water.
- Flocks of birds circle water.
- Game trails heading downhill may lead to water. Where two game trails converge, follow them from the point where they join.
- Grazing animals (cows, deer, etc.) are usually close to water, unless they're migrating.
- Human tracks may lead to a well, bore-hole, or soak. Replace any covers after use to prevent evaporation and contamination.
- Swarming insects indicate water close by.

Large fish have water along their spine and in their eyes. To get it, carefully cut a fish in half to get the fluid along the spine. Suck the eye as well. The rest of the fluids in fish are rich in protein and fat. They're good to consume if you have excess water to digest them with.

Snow and Ice

Most ice is safe to drink. Melt it first to prevent hypothermia and the need to use any excess energy.

Fresh snow is fairly safe, but purify any snow more than a few hours old.

It's better to melt ice than snow. It uses up less energy and has a higher water-per-volume ratio.

In arctic waters, old sea ice has less salt than new ice. Old sea ice is bluish, has rounded corners, and splinters easily. New sea ice is gray, milky, hard, and salty. Water from icebergs is fresh but dangerous to collect.

You can melt ice (or snow) using your body heat as you travel. Put the ice in a container and place the container between two layers of your clothing. Never put it directly on your skin, and keep it away from any major arteries.

Another way to melt ice is with a fire. Place a container with a little bit of ice near (not in) the fire. Stir it frequently and add more ice as it melts. Speed up the process by placing clean, hot rocks or water in the container.

Once it's melted, keep your water close to you so it doesn't freeze again. Keep a little room for it to move in your container, and slosh it around occasionally to slow the freezing process.

Nuclear Situations

In a nuclear situation, procure water from the following sources, in order of preference:

- Underground sources such as springs and wells.
- Water from pipes/containers in abandoned buildings.
- Snow. Dig six or more inches below where the surface was during the fallout.
- Streams and rivers.
- Other open water sources such as lakes, ponds, and pools.

No matter where you get it from, purify it.

Related Chapters:

- Insects

WATER FILTRATION

Although not as effective as commercial water filters, an improvised filter can clean water enough for you to drink it, provided it isn't too contaminated. When you have the resources, purify the water too.

Here are some ways to filter water:

Layer Filter

In a layer filter, the water passes through layers, from coarse to fine, to remove contaminants. What you use for each layer depends on what you have. The more different layers you have the better.

To make a layer filter from a plastic bottle, cut the bottom off and turn it upside down. Place a layer of cloth on the bottom, which was the nozzle. This prevents the other layers from falling out.

Fill it up with materials, from fine to coarse. Charcoal is a great bottom layer because it can absorb chemicals. Even charcoal (not ash) from your campfire will work, although not as well as activated charcoal.

The complete layer system may look something like this:

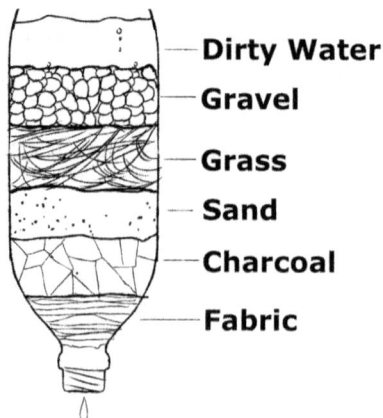

— **Dirty Water**
— **Gravel**
— **Grass**
— **Sand**
— **Charcoal**
— **Fabric**

Sapwood is a good bottom layer instead of or as well as charcoal. It's slower to filter, but removes more micro-organisms. If you "cork" the nozzle of a bottle with it, expect four liters of drinking water a day. The bottle must be sealed tightly.

If you don't have a bottle, you can construct a tripod and use several pieces of cloth to hold each layer.

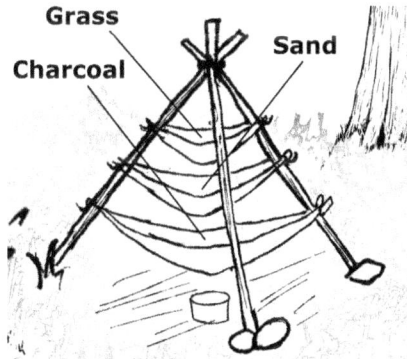

Digging and Sedimentation

This method won't remove any micro-organisms, but is better than nothing when collecting water from a slow or still water source, such as a swamp, lake, or pond.

Dig a hole 3m (9ft) inland from the swamp's edge. It must be deeper than the water table. The soil will filter the water as it seeps in. Strain the water through material if you have it. If not, let it sit for at least half an hour.

Letting it sit is the sedimentation process. You can do this with any water to give heavier particles time to sink to the bottom. Once they've settled, scoop the cleaner water off the top.

WATER PURIFICATION

There are several ways to purify water. Some are better than others, but which one you use will depend on what resources you have.

Most of these methods will kill pathogens, but only distillation will make seawater, urine, or chemically tainted water drinkable.

Boiling Water

Boiling water is easy to do, and is one of the most effective forms of purification.

As a rule, rapid-boil the water for one minute at sea level, and one additional minute for every additional 300m above that.

When you don't know your elevation, a three-minute rapid-boil is a safe bet, unless you're in the mountains. The longer the boil, the safer it is, but the more water will evaporate.

If you don't have a fireproof container, use natural substances like bark, bamboo, or any non-toxic plant. Heat dry rocks in or next to a fire and drop them into the container of water until it boils. Don't heat wet, porous, slate, or softer rocks, as they may explode.

You can use the hot-rock method with a hard plastic container too, but it's a last resort, since plastic is toxic.

Pasteurizing

When you are unable to bring water to a rapid boil, heat it as much as possible for 20+ minutes. Get it hot enough to burn you if you touch it.

SODIS

The SODIS method uses UV rays to kill the pathogens. It is effective and easy, but takes a minimum of six hours.

To use it, you need a transparent plastic bottle (PET), such as a soda bottle, and the sun. The bottle must be 2L or smaller, clean, and with minimal damage.

Your water must be as clear as possible before you out it in the bottle. If it isn't, filter it first. To test if your water is clear enough, put it in the bottle. If you can count your fingers on the other side while looking through the bottle, it's okay to use.

Fill the bottle 3/4 full of water and seal it. Shake it vigorously for 20 seconds and then fill it up to the top. Leave it in direct sunlight. How long it needs to sit outside depends on how much sun there is:

- Sunny = Six hours.
- Partly cloudy = A full day.
- Very cloudy = Two days.
- Raining = It won't work, but you can drink the rainwater instead.

For best results, place the bottle on a reflective surface, such as aluminum foil or a metal sheet, and slope it towards the sun.

When you don't have a suitable bottle, any container will work, but only with a thin layer of water—that is, one a maximum of 15cm (5in) deep.

Chemical Purification

There are a few options for purifying water with chemicals.

Purification tablets are made specifically to purify water. Use them as per the manufacturer's instructions. If you don't have instructions, follow these guidelines:

- Use one tablet per liter of clear water.
- Use two tablets per liter of cloudy or very cold water.
- Shake/stir well, and wait at least 30 minutes.

A 2% tincture of iodine is a common chemical in first aid kits. Guidelines for water purification are as follows:

- Use 5 drops per liter of clear water.
- Use 10 drops per liter of cloudy or very cold water.
- Shake/stir well, and wait at least 30 minutes.

Bleach is a common household chemical that you can use to purify water. Make sure it has no additives, such as scents. A solution with 5.25% calcium hypochlorite is good.

- Use two drops per liter of water.
- Shake/stir well.
- Wait at least 30 minutes for clear water.
- Wait at least 60 minutes for cloudy or very cold water.

Potassium permanganate is another chemical you might find in a first aid kit or medicine cabinet. It's commonly used to help with skin conditions. To use it for purifying water, add 0.1g of potassium permanganate crystals (about three crystals) for every liter of water, and stir. The water should become light pink. If it's any darker, don't drink it.

Note: Using chemicals to purify water is a temporary solution. Long term use may damage your health.

WATER DISTILLATION

Distillation can purify almost anything that contains water, including mud, saltwater, and urine No other filtration or purification is needed.

The basic process is to boil the tainted water and collect the steam. When the steam cools, it turns into drinking water. There are many ways to do this. Here are a few ideas:

Fabric Distiller

This combines the soak-and-squeeze method of water collection with boiling water.

Boil the water and suspend any absorbent material over it to collect the steam. Wring it out to get the fresh water.

Adapt this for the digging and sedimentation water filtration methods by placing hot rocks in the hole to boil the water.

Pot, Lid, and Cup

Tie a cup to the handle of a pot lid so that it hangs the right way up when the lid is put on the pot upside down.

Fill half the pot with water and put the lid on upside down. The cup shouldn't dangle into the water. Boil the water. The distilled water will drip into the cup.

Pot and Lid

When you don't have a cup or string, put the cover on with the handle facing up. Angle it so the condensed steam flows into your container.

Open Container

Use this method when you have a container, but no lid.

Make a small plastic "tent" on top of your container. Boil the water and collect what runs down the plastic.

Tube Distiller

This method minimizes waste, but is more complicated to use.

Put water in a container and seal it with plastic. Stick a piece of tubing that leads to a collection container in the top of the first container. When you boil the water, the steam will be forced up the tube, and will drip into the collection container.

Solar Still

A solar still combines condensation water collection with solar distillation. It's a viable way to produce water in the desert if you have the materials. To produce enough, you'll need at least three of them per person.

First, you need to choose a good site. Look for one with the following characteristics:

- A good possibility that the soil will contain moisture (a dry stream bed or low spot, for example).
- A lot of sunlight.
- Ground that's easy to dig in.

Dig a hole with sloped sides and put a container in the middle of it. If you have contaminated water you want distill, put your collection container inside the container with the contaminated water.

Put plants in the hole as an extra moisture source (if available). Place them on the sloped sides of the hole.

If you have a tube to use as a drinking straw, run it from your collection container to the outside of the hole. Cover the hole with a plastic sheet (preferably a clear one) and secure it with rocks or dirt.

Place a stone in the middle of the sheet so that the condensation will drip into your container.

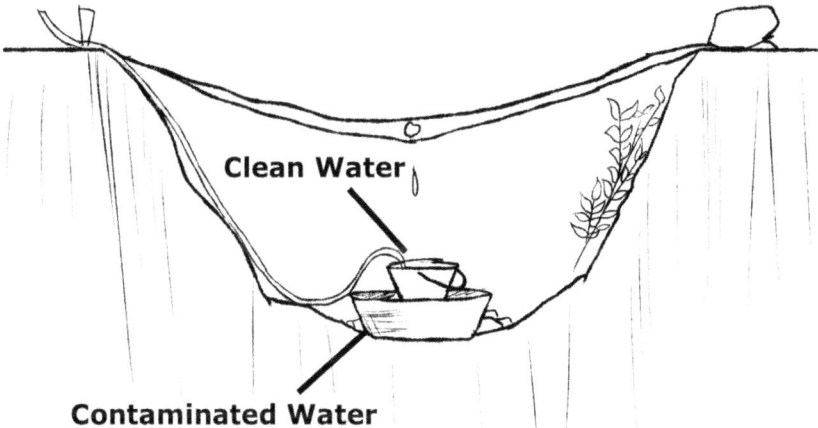

If you only have one container, but want to distill contaminated water, dig a hole 25cm (10in) from the still's edge. Make it 25cm (10in) deep and 10cm (4in) wide. Pour the polluted water in the hole. Don't spill any of it near the plastic. The earth will filter the water into the solar still, and then the solar still will distill it.

Mini-distiller

Make a mini solar distiller out of a plastic bottle (with the lid) and soda can (or similar items). Cut off the bottom off the plastic bottle and the top off the can. Create a water collection lip out of the bottle by folding the bottom of it up into itself.

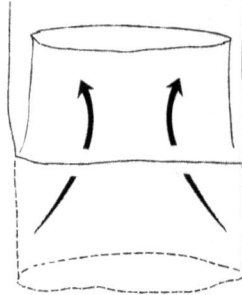

Fill the can up with the contaminated water and set it in the sun. Put the plastic bottle over it. Ensure the lid is sealed tightly.

Drink the water straight out of the bottle. Be careful not to spill it.

Related Chapters:

- Water Filtration

FOOD

While on the run, you'll need to keep your energy up, but you won't want to spend too much time acquiring food—at least not until you've created enough distance between you and your enemy.

Store up food while in captivity, but eat off the land whenever possible. Forage for food you can eat raw (edible plants and insects) and conserve your rations for when they're needed, such as when there's no other food available or when your enemy is too close for you to stop.

Once you create enough distance, you can catch fish and/or small animals, such as birds or reptiles. Use the lessons in the stealth movement chapters to get close to your prey.

Hunting or trapping larger game takes too much time, and preparing game leaves significant signs of presence.

COOKING

Cooking food is important for killing bacteria and parasites, but fire is a big sign of presence. The *Fire* section will teach you how to minimize this.

The best way to cook is by boiling. It is the most likely way to kill harmful organisms, and retains the most nutrients. It is also the most palatable way to eat "strange" foods such as insects, allowing you to pulverize them and cook them in a soup or stew.

To get the most nutritional benefit, drink the water you boil the food in. The exception to this is if you are boiling out toxins from plants or other materials.

The downside of boiling is that you need water. Some alternative ways to cook are:

- Roasting on a skewer.
- Frying on a thin flat rock, sheet of metal, etc.
- Using a solar oven. This takes a long time, but negates the need for a fire. Improvise one with aluminum foil or a survival blanket.

Once you have hot coals, use them as a stove. Flatten the fire, pile the embers, and then compact them. Place your pot directly on the coals. When time is limited, cook over the flames. You can make a simple potholder out of sticks.

Drying or smoking food is a good way to save it for later. Suitable foods for this include fruits, nuts, and thin strips of meat and fish.

Clean and cut what you want to dry. The thinner you slice your food, the faster it will dry. Remove all the fat from meat.

Sun drying takes too long for the evasive survivor. Place the strips near the fire instead, either on hot rocks or by hanging it. Ensure there are no folds; otherwise, there will be moist spots and it will grow bacteria.

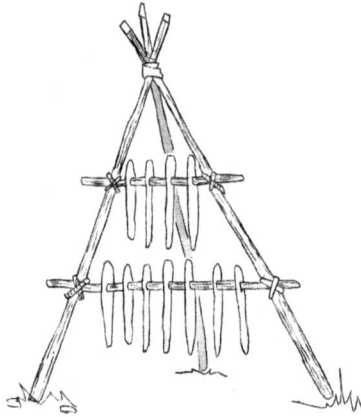

To test if it's dry enough, bend it. When it cracks, it's ready.

At night, wrap the food up to prevent critters and moisture from the morning dew from spoiling it.

EDIBLE PLANTS

Raw plants are sometimes ideal sources of food for the evasive survivor, but many are unsafe to eat.

To ensure your health, get an edible plants field guide for the area you plan to be in, and learn to identify those you can eat raw. Practice identifying and preparing those you need to cook. When collecting plants to eat later, try not to crush them.

You can use the edibility test for unknown plants. It's risky and takes time, but is useful when you're desperate.

Signs of Edible Plants

Any plant with one or more of the following characteristics is a good candidate for further testing:

- Animals eat it.
- It has blue or black berries.
- There are five petals at the end of a single fruit (rose family). The fruit and the flowers are likely edible.
- It resembles plants humans cultivate.
- It has segmented fruits and berries, like raspberries or blackberries.
- There are single fruits on a stem.
- It's a seed from a cone-bearing tree.
- It's a seed from a type of grass.

Plants growing in water or moist soil are usually the most palatable.

Most types of seaweed are edible raw and have high nutritional value. Wash them in fresh water before consumption. Eating too much seaweed can have a laxative effect.

Signs of Inedible Plants

Plants with any of the following characteristics are not edible unless positively identified:

- Bulbs.
- Fruit divided into five segments.
- Grain heads with pink, purplish, or black spurs.
- Legumes, including beans, peas, and seeds in pods.
- Mature bracken.
- Milky sap.
- Mushrooms and fungi. Mushrooms and fungi must be positively identified. The edibility test is not reliable.
- Age or wilting.
- Skin irritants.
- Red.
- Shiny leaves.
- Slime.
- A smell of bitter almonds or peaches. Crush a small portion and smell it.
- Strong acid odors.
- Three-leafed or whorled-leafed growth patterns.
- Tiny barbs, spines, thorns, or fine hairs.
- Umbrella-shaped flowers. Common exceptions include carrots, celery, dill, and parsley.
- Signs of being worm-eaten.

If the plant has none of the above or you think you have positively identified it as safe, move on to the edibility test.

Edibility Test

The edibility test is time-consuming, so make sure there is a good supply of the plant you want to test.

Ideally, you want to test it by boiling first. If it's edible after boiling, test it again raw. If it's is edible raw, it will be okay to cook by any method.

First, select one part of the plant to test, such as the leaves, flowers, stems, or roots. Just because one part of the plant is edible doesn't mean all of it is, so test each part separately. Wash it with drinking water.

Crush the plant part in your fingers and rub it on your inner forearm. Wait 15 minutes. If there's any irritation, discard it.

If it passes the skin test, test it for consumption. You must fast for at least eight hours before and after. You can drink water as long as you know it's safe.

Keep some charcoal mixed with water close by to soak up toxins. If you have any bad reaction (burning, bitterness, nauseating taste, itchiness, swelling, etc.) at any stage, discard the plant immediately.

- Boil the plant part, then touch a small portion of it to the outer surface of your lip. Wait three minutes.
- Place a bit on your tongue and hold it there for 15 minutes.
- Chew it and hold it in your mouth for another 15 minutes. Do not swallow it.
- Swallow it and wait eight hours.

If you experience a bad reaction:

- Drink lots of hot water
- Do not eat until the pain subsides.
- If the reaction is severe, induce vomiting, then consume the charcoal-and-water mixture.

If there's no bad reaction:

- Eat half a cup of plant prepared the same way and wait another eight hours.

If there's still no bad reaction, consider it safe to eat when boiled. Repeat the test without boiling it if you want to eat it raw.

Additional Notes

If a plant is found to be edible, eat it only in moderation, even after testing.

If the plant's properties change during the year, or you wish to prepare it a different way, you must repeat the test.

It's best to cook all underground portions (roots) of plants, even if they pass the edibility test raw.

Plants that are most likely to be safe to eat raw are leafy green plants, ripe fruits, berries, and nuts.

The same part or plant may produce varying reactions in different individuals.

Plant Identification

These plant identification techniques will make it easier to record, remember, and identify different plants.

Record detailed descriptions of all plants you test for edibility: what they look like, smell like, taste like, how you prepared them, where you found them, what season they were in, etc. Draw pictures.

There are many more plant identifiers than those listed here, but committing the basics to memory is helpful since you probably won't have a guide handy.

Smooth   Toothed   Lobed

Basic Leaf Margins

Egg   Elliptical   Fingered

Heart   Lanced   Oval

Rectangular   Star   Triangular

Basic Leaf Shapes

Simple

Opposite

Alternate

Whorled

Compound

Basal Rosette

Basic Leaf Arrangements

Bulb

Clove

Corm

Crown

Rhizome

Taproot

Tuber

Basic Root Structures

- Bulbs, such as onions, show rings when cut in half.
- Cloves, such as garlic, separate into smaller segments.
- Corms, such as taro, are like bulbs, but are solid when cut rather than possessing rings.
- Crown roots, such as asparagus, look like a bunch of stringy hairs.
- Rhizomes, such as ginger, are networks of plant roots, and usually extend horizontally from the main root of the plant.
- Taproots, such as carrots, usually grow as one per root.
- Tubers, such as potatoes, are found in clusters or on "strings" under the plant.

More Information

Plants have many other uses, such as providing material for cord or herbal medicines.

Search the internet for the area you plan to be in. Use search strings like:

"Edible plants of [country]"

or

"Medicinal plants of (enter country)"

There are several free to read plant guides available online. My favorite one is *Edible and Medicinal Plants*:

https://docs.google.com/file/d/0B6GE42-kvADvNmE4MDBmMzAtMDU3NC00NWZiLThhY2QtMmYwNWRmNjZkNWQ0/

Another good one that is not strictly about plants is *A Complete Handbook of Nature Cures*:

https://docs.google.com/file/d/0B6GE42-kvADvYjEyOTQxM2EtZjBkMi00Njg1LWFjYWEtMmU5ODg0MjRhYzEz

There are also some useful apps you can download to your phone, a few of which have photo recognition.

Apologies if the Google Doc links above are no longer working. At the time of writing, they are, but I have no control over whether the original uploader decides to take them down.

Related Chapters:

- Cord

INSECTS

Many people will find eating insects "strange," but as a survival food, they are highly nutritious and easy to find in most places. You can eat most of them raw if you need to, but it's best to have a guide.

Look for insects in damp and dark places, such as in wood, under rocks, or underground. Beware of dangerous animals that like the same spots, such as spiders, scorpions, and snakes.

Insects that live underground, such as earthworms, will surface after hard and/or prolonged rain.

Avoid the following insects:

- Anything that stings or bites.
- Caterpillars. Brush hairy caterpillars away in the direction they are traveling.
- Dead or sick insects. Do not save dead insects.
- Disease carriers such as ticks, flies, mosquitoes, roaches, etc.
- Those that feeding on waste.
- Those found on the undersides of leaves.
- Those that are hairy or brightly colored, especially if they're orange and black.
- Those that have a strong smell.
- Those that produce a rash when touched.
- Those that move slowly in the open.
- Spiders.

You can use insects you don't eat as fish bait.

Although most edible insects are safe to eat raw, cooking them will make sure of it and will also improve the flavor, especially if they're larger than a grasshopper. Any easy way to do this is to crush them into a paste (or powder if they're dried) and boil them in a soup. Remove the wings and legs from larger insects, and take the armor off beetles.

Here are some details on specific insects:

Bees

Consider carefully whether to attack a beehive. It won't be worth it if you're hiding from your enemy. Avoid wasps and hornets.

Follow the bees back to their hive. Return at night and smoke the hive using a grass torch. Seal the hole to kill the bees.

Eat their honey and the honeycomb. To eat the bees, remove their wings, legs, and stingers. Use the wax for waterproofing material or making candles.

Crickets, Grasshoppers, and Locusts

Find these insects early in the morning near the tops of tall plants.

Use a piece of clothing or a leafy branch to swat them. Try not to squash them. Remove their antennae, legs, and wings.

Snails and Slugs

Stay away from sea snails and/or those have brightly colored shells.

For other snails and slugs, starve them for a few days or feed them safe greens. They may have eaten plants that are poisonous to humans, so you need to wait for them to excrete those toxins.

Boil them for 10 minutes.

Termites

Break off chunks of termite mounds and dunk them in water. Another (but much slower) way to get the termites is to insert a twig into the mound and then gently withdraw it. The termites will hang on to it.

Remove their wings before cooking. You can also eat their eggs.

A piece of the termite nest in the fire (or on the coals) is a good mosquito repellant.

Worms

Either starve them for a day, squeeze them between your fingers to clear out the muck, or put them in potable water for at least 30 minutes.

Eat them raw or cook them.

Related Chapters:

- Clothing
- Cooking

WATER FORAGING

This chapter covers water-based animals that you can forage for. As a rule, boil shell foods for at least five minutes, and preferably 10.

Shellfish

Many shellfish are not safe to eat. Avoid the following:

- Anything in polluted areas.
- Those with cone-shaped shells.
- Shellfish found above the high-tide mark.

Find mollusks in shallow, fresh water with a muddy or sandy bottom. Look for:

- The narrow trails they leave in the mud.
- Their dark elliptical slit.

Near the sea, look in the tidal pools and the wet sand.

Only eat mollusks collected live. Steam, boil, or bake them in their shells.

Catch clams and shellfish at low tide on tidal flats, tidal pools, harbor/bay sandbars, etc. Shellfish cling to rocks along beaches or extending as reefs into deeper water.

Bivalves

Bivalves are aquatic mollusks with hinged shells, such as clams, oysters, mussels, and scallops. Edible bivalves will close tight when tapped.

Mussels are poisonous in tropical zones during the summer. Black mussels are always poisonous in the Arctic.

Look for mussel colonies in rock pools, on logs, or at the bases of boulders.

Gastropods

Gastropods are another type of mollusk. Abalone, conches, and other sea snails fall into this category. They have trapdoor entrances to their shells, which should close tightly if the shells are shaken.

Limpets and abalones anchor to rocks. Pry them off with a knife. They should be difficult to dislodge. If they aren't, don't eat them.

Sea Cucumbers

Find sea cucumbers on a seabed or on sand. Collect them live and boil them for at least five minutes.

Sea Urchins

Sea urchins cling to rocks just below the low-water mark. Only collect the ones whose spines move when they're touched. Boil them, split them open, and eat their insides. Do not eat sea urchins if they smell bad when opened.

Crustaceans

Crayfish are active at night. During the day, look under and around stones in streams. Scoop one gloved hand behind a crayfish while scaring it backwards with the other hand.

Alternatively, tie bits of offal to a string. When the crayfish grabs the bait, pull it to shore before it has a chance to let go.

Find saltwater lobsters, crabs, and shrimp anywhere between the surf's edge and 10m deep in the water. Shrimp may come to a light at night, enabling you to scoop them up with a net.

Look for freshwater shrimp in floating algae or in the muddy bottoms of ponds and lakes.

Catch lobsters and crabs at night with a baited trap or a baited hook. Crabs will come to bait placed at the edge of the surf. Trap or net them.

Related Chapters:

- Fish

FISH

When you're close to a body of water, fish are a good source of food. Catching them is faster and easier than hunting game, and doesn't leave a large a sign of presence, unless you're careless. It still takes time, though, and you risk being spotted.

Do not eat a fish that is or has:

- Dead and/or floating, unless you just killed it.
- Flabby skin.
- Flesh that remains dented when pressed.
- In reefs or lagoons, especially in the tropics. Instead, fish from the reef on the seaward side of the lagoon.
- A non-existent or single small pelvic fin.
- Pale, shiny gills.
- A parrot-like mouths.
- A round body with a hard, shell-like skins covered in bony plates and/or spines.
- Small gill openings.
- A slimy body.
- Spines.
- Sunken eyes.
- An unpleasant odor. Fish from the ocean will have a clean fishy smell, which is a sign they're okay to eat.

The above signs mean the fish is either poisonous or has spoiled. Cooking it does not help.

When fishing, be careful of:

- Fish that have teeth or spines.
- Handling fish with sharp fins and gills. Use a cloth or gloves.
- Large fish, especially when you're in a raft. It is better to catch small fish than risk capsizing or injury.

- Sharks. Don't fish while they're around and always be on the lookout for them, even in shallow water. There may be larger predatory fish about in places where lots of fish are darting.
- Sea snakes. They are edible but poisonous.
- Underwater crevices. Don't put your hands in them.
- Walking in water. Stir up the bottom in front of you to uncover camouflaged creatures.

Finding Fish

Schools of fish give you the best chance of success. Look for ones jumping out of the water, and/or for circular ripples in the water.

During the day, look in shady/sheltered places with less current, such as in deep pools, under overhanging brush, around submerged items (banks, rocks, logs, foliage, etc.), and under your raft.

On most coasts, the best time to fish from the shore is roughly two hours after high tide come in.

When a storm is approaching, it's a good time to fish. After heavy rain, it's difficult.

In cold weather, fish prefer shallow water in the sun. When ice is on a lake, they will be deeper.

When a river is in flood, fish in slack water—on the outside of a bend or in a small tributary where it enters the main stream, for example.

Attracting Fish

Attract fish with the foods they are used to eating, such as berries that overhang their water or insects that breed in it. Examine the stomach contents of your first catch for more clues.

Scatter bait across the water and use the same bait on your hook or in your trap.

Use the following as generic baits/lures:

- Anything shiny drawn through the water, like coins or tin.
- Feathers tied to a hook with thread to simulate a fly.
- Leftover parts of fish you eat, like intestines, eyes, or dorsal fins.
- Live bait such as worms, insects, maggots, or small fish.

A flashlight held above water at night will attract fish. A similar method is to lay a mirror or flat piece of metal on the water to reflect the moonlight. This also works with octopus.

Simple Barrier Trap

When a stream widens out a little, you can get fish to gather by building a barrier. It's not really a trap, but they're easier to catch when concentrated.

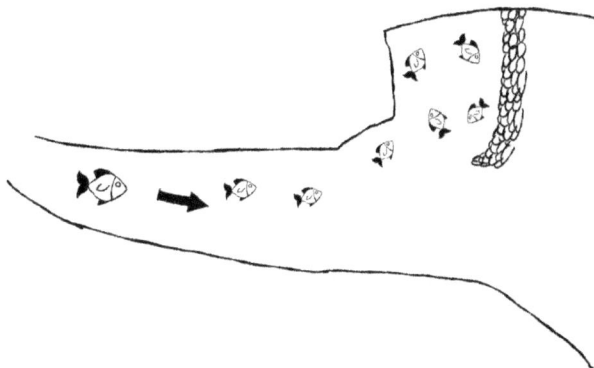

Weir

The weir trap is simple to construct out of easy-to-find materials like sticks.

Push sticks into the mud in a half box shape, with the opening facing upstream. Put the sticks close together.

Create a funnel into the opening using more sticks. Place it so that the bottom of the funnel goes inside the boxed area. Ensure the bottom of the funnel is big enough to allow larger fish to swim in.

Put bait in the box area and/or stir up the mud outside the opening of the funnel so the fish swim to the clearer water inside your trap.

The fish can swim/be driven in, but have a difficult time finding their way out. If you want to keep fish alive in there, you can block the entrance once you've caught them.

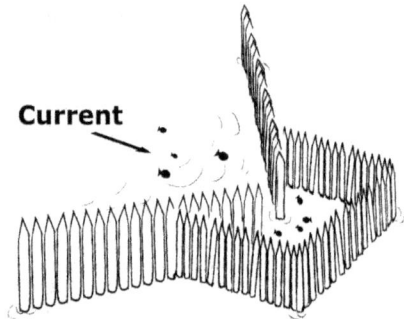

Bottle Trap

This trap works on the same principle as the weir trap, but uses a plastic bottle. It's only good for catching small fish.

Cut the top off a plastic bottle, just below the neck. Put some bait inside the bottom part of the bottle, then insert the nozzle into the bottom piece of the bottle. If necessary, put a weight inside it so it doesn't float away.

Similar traps can be made from hollow logs and/or twigs.

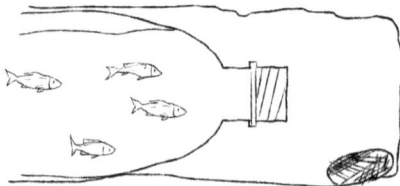

Fish Net

Constructing a net with cord is possible, but not practical for the evasive survivor. Instead, use any material large enough to scoop up fish. You can attach it to a stick to extend your reach.

Angling

This is fishing as most people know it. In a survival fishing situation, it's all about improvising fishing tackle.

Any thin cord, such as the dental floss or the inner strands of paracord, can be used as fishing line. Thread from your clothes is too thin, and a regular shoelace is probably too thick, though you can try it. Making cord from plants is another option.

You can improvise hooks from bone, nails, wire, pins, thorns, wood, etc. Small hooks are better than large ones because they can catch big fish too, but you need to ensure they're not too flimsy.

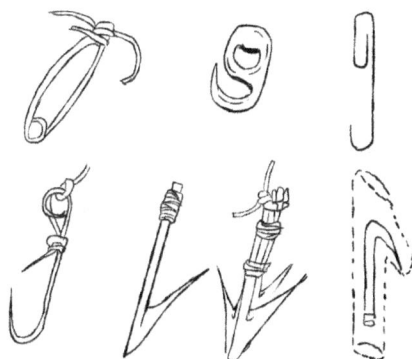

Angling Tips

A fishing rod is better than a hand line. Tie your hook to one end of the line and the other to a stick (your fishing rod). The size of the stick depends on the location you want to fish from. Don't use anything brittle, or it may break when a fish tugs it. Ensure the knots are secure.

Tying a float above the hook will show when you have a bite. It will also help to control how deep the line will sit.

An improvised sinker (e.g., a stone) between the float and the hook will stop the line from trailing along the water or too near the surface, while leaving the hook itself in movement. To get a deeper hook position, place the sinker at the end of the line and tie the hook somewhere above it.

Put your baited line in where there are fish. Ensure your shadow is not casting over your fishing area.

Once a fish bites, scoop it out of the water with an improvised net as opposed to lifting it out with your line. This prevents your improvised line from breaking. Using a spear to retrieve it is another option.

When fishing in salt water, be careful of handling your fishing line with bare hands. The salt will give it a sharp cutting edge.

If you're at sea, kill fish before bringing them into your raft, and be very careful not to puncture your raft with hooks or other sharp instruments.

Gorges

A gorge is a small shaft of wood, bone, metal, or other material. It is sharp on both ends and notched in the middle where you tie it to your fishing line. Bait the gorge by placing a piece of bait on it lengthwise.

When the fish swallows the bait, it will also swallow the gorge, which will lodge in the fish's throat. Once the fish has swallowed the bait, and not before, jerk the line to catch it.

Gorges work well on eels and catfish, as they swallow without biting.

Set Lines

A set line is a mix between a trap and angling. It uses a line and hook, like angling, but you can leave it unattended while you sleep/do other tasks. Check them before first light, so other fish don't eat your catch before you get to it.

To make a set line, have one main line and attach baited lines and hooks at different intervals. Make sure they will not slip along the main line or get tangled up with each other.

Anchor one end of the main line to something on the bank. Weight the other end of the main line and place it in the water so it is taut. If you need to, readjust your line and hooks so they hang free in the water.

If you find that you catch many fish at a certain spot on the line, put more hooks near that spot.

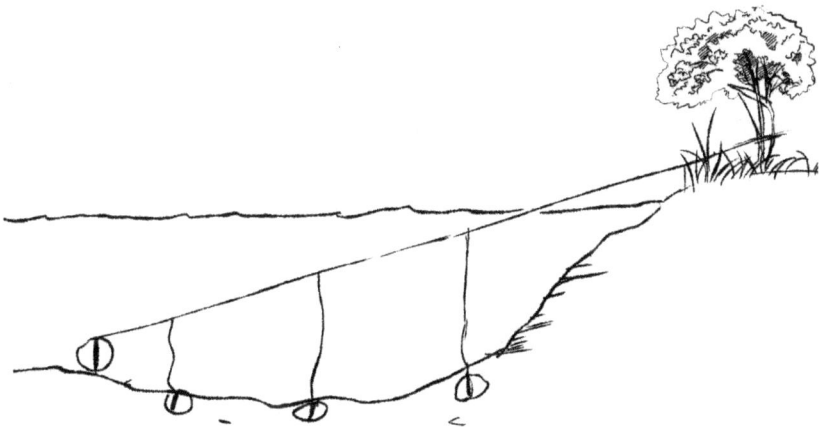

Stakeout

A stakeout is a covert version of set lines.

Drive two supple saplings into ground where you want to fish so their tops are below the surface. Tie your set line between them.

Tippet

A tippet is a set-line setup for ice fishing.

Whatever you use for an anchor, ensure it's big enough not to fall through the hole. Secure it in a snow mound.

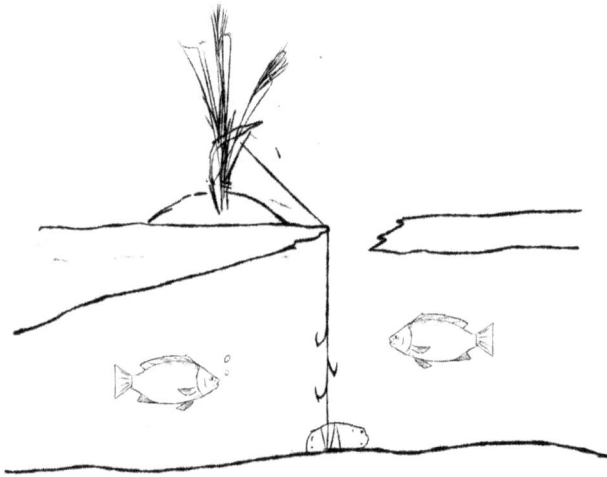

Snagging

When you can see fish but they aren't taking bait, try snagging.

Tie several hooks to a pole and lower it into the water. Suspend a bright object 20cm (8in) above the pole. Pull up sharply to catch fish when they come to inspect the object.

Spear Fishing

Spear fishing is most effective in or near shallow water (about waist deep) where there are a decent number of medium to large fish.

You can also use spears to lift fish out of the water when angling or trapping.

To make a basic spear, get a long, straight sapling and tie a knife to it. Make sure the knife is secure, so you don't lose it. When at sea,

use an oar instead of a stick

Alternatively, you can sharpen the tip and fire harden it.

To fire-harden wood, hold it over a bed of hot coals. Rotate it slowly. It will begin to hiss/steam. Do this until it is a light brown. Don't char it.

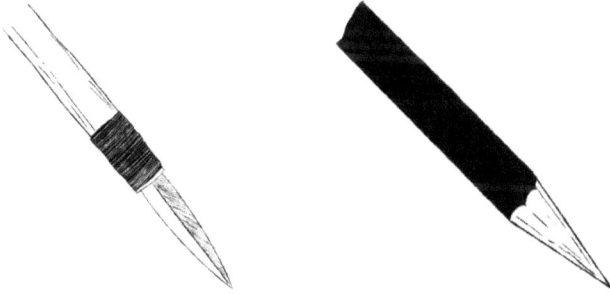

Single-point spears work, but the best spears for fishing are multi-pronged.

To make a two-pronged spear from one stick, split it down the middle about 15cm (6in). Wrap some cord directly below the split to prevent it from splitting more. Do this as tightly as you can. Place a small piece of wood in the split to separate the prongs, then sharpen both sides into points.

Make a multi-pronged spear from bamboo in the same way.

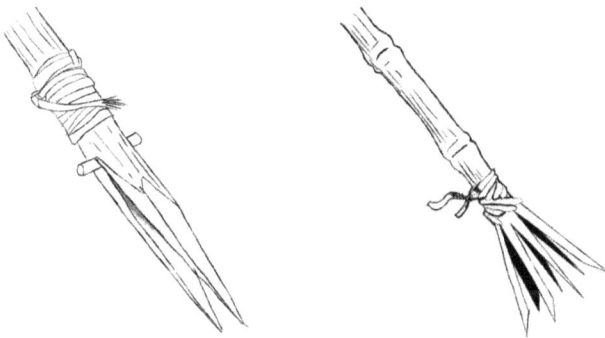

To use a spear, thrust it as opposed to throwing it.

Stand so your shadow is behind you, and sprinkle some bait in front of you. Stay still.

Place the spear point in the water and move it slowly towards the fish. Aim slightly below the fish to allow for refraction. Practice stabbing at something in the water beforehand to get used to it.

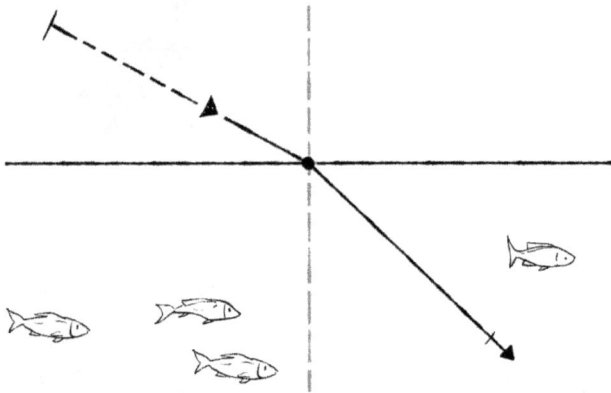

Use a swift motion to impale the fish. Drive it all the way to the bottom of the water if possible.

Hold the spear with one hand and grab the fish with the other (or use a net), as opposed to lifting the fish out with the spear.

Fish Narcotics

There are certain plants you can use to stun or kill fish. Parts of plants that contain rotenone are ideal. Crush the right part and throw it into the water. This works best in slack, warm waters. The colder the water, the longer it takes. It won't work in water below 10C (50F).

Make sure you collect all the dead fish so there is no sign of presence downstream.

Wash the fish before consuming it to remove any traces of the poison. Although rotenone is generally safe for humans, some of these plants have other poisons in them.

Here are some examples of fish narcotics. Research the area you are planning to go to for specifics.

- Indian berry seeds (*Anamirta Cocculus*). SE Asia and India.
- Purging croton seeds (Croton Tiglium). Asia. These seeds are also good for combating constipation, detoxing the body, and other medicinal purposes.
- Fish poison tree seeds and bark (Barringtonia Asiatica). Tropical Regions
- Derris powder roots (Derris elliptica). Grind them into a powder and mix it with water. Throw a lot into the water. SE Asia and SE Pacific.
- Hoary pea leaves and stem (Tephrosia). Tropical and warm-temperate regions.
- Desert rose (Adenium Obesum). Deserts of southern and eastern Africa and Arabia. Be extra-careful with this plant, as the roots and seeds contain a deadly poison. Instead of crushing the plant, splash the branches into the water and catch the dizzy fish.
- Doveplant (Croton Setigerus). Western North America and some parts of Australia.
- Soap-plant bulbs (Chlorogalum). Western North America.
- Walnut's green husk (Juglans Cinerea or Juglans Nigra). Eastern North America.

Lime is also a fish narcotic. Burning coral, seashells, and unoccupied rainforest snail shells will create lime. Throw it into the water.

Preparing Fish

Eat small fish whole. Bleed and gut anything over 5cm (2in) soon after catching it. Cut its throat to bleed it, and remove the gills. Slice across the underside of it, starting at the anus and working towards its gills. Remove the entrails, but keep the roe, which runs down side of fish. Clean it thoroughly.

Scaling is not necessary, but if you want to do it, draw the knife from the fish's tail to its head. Even if you scale a fish, it is a waste not to eat the skin.

To skin eels and catfish, pass a stake through them and fix it across two uprights. Cut the skin away and draw it down towards the tail.

Cooking

All freshwater fish is fine to eat once it's cooked. You can eat salt-water fish raw if it's been caught far out at sea.

Boil or stew fish for at least 20 minutes, especially if it's been caught in slow-moving water.

Bake fish by burying it in the dirt under your fire. Wrap it in leaves or pack it in mud before burying it.

Dried fish will stay edible for several days.

When cooking shark, cut it into small cubes and soak it overnight in fresh water. Boil it in several changes of water to get rid of the ammonia flavor.

To cook octopus, boil the body and roast the tentacles.

Related Chapters:

- Cord
- Cooking
- Insects

BIRDS

All birds and their eggs are edible, except for the Little Shrikethrush and the Pitohui. Both of those are found in New Guinea.

These methods for catching birds will also work for bats and similar flying animals.

Bird Eggs

Before attempting to catch or hunt birds, it's worth looking for their eggs. They are good to eat and much easier to get.

Boiling an egg is best. If there's an embryo inside, you can remove and roast it on an open flame.

When boiling isn't possible, roast it. You must first pierce the yolk; otherwise, it will explode in the fire. Create a hole as small as you can in the flat end of the egg, and then stick something down it. Place the egg near the fire open end up, and turn it after a few minutes so the heat is spread evenly.

Where to Find Birds

Observe the direction of their flight in the early morning and late afternoon, which will lead you to feeding, watering, and roosting areas. Bird calls are an indication of nesting areas. Droppings can indicate a night roost.

Baited Hooks

A good way to catch seabirds is with a fishing hook and line. Wrap the hook with bait (fruit, fish, etc.) and throw it in the air for the bird to catch. A variation of this is to wrap food around a stone instead of a hook. The stone in the bird will cause it to crash.

Bola

A bola is an easy-to-make weapon to use against a low-flying flock.

To make a bola, you need three lengths of cord 50cm (20in) in length and three rocks 1/4kg (1/2lb) in weight. Join the lengths of cord together at one end and tie a rock onto each other end.

To use the bola, hold it by the knot and swing it above your head in a circle. Let go of the knot so it flies at your target.

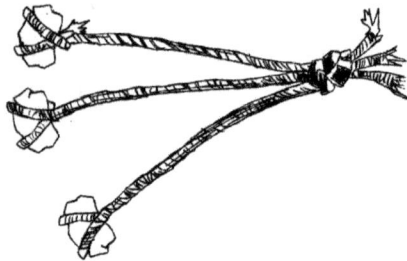

Rock Sling

To make a sling, you need two pieces of cord 50cm (20in) in length and a small rectangular piece of material about the size of your palm. Soft leather is best, but any cloth that's not too flimsy will work.

Punch two holes in the cloth, one on each of the shorter sides. Attach one piece of cord to each hole. On the other side of the cordage, tie a small knot in one end and create a finger loop in the other.

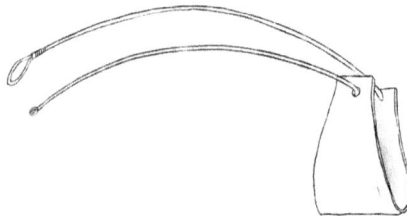

Use larger pebbles as ammunition. Try to find ones that are 2.5cm (1in) in diameter each.

Put your middle finger through the finger loop and hold the other end with your thumb and pointer finger. Swing the sling vertically in a forward circle at the side of your body.

Point at your target for increased accuracy and release the knotted cord.

Using a sling or bola effectively is not hard, but it takes a little practice to get your aim right.

Clubbing

Use a stealth crawl to approach ground colonies, and throw a stone or rabbit stick to knock one out.

At sea, you may be able to entice seagulls with bait. Once they're close enough, club them with an oar.

Preparing Birds

Prepare a bird soon after you kill it.

Stretch the neck, cut the throat, and hang it upside down to drain the blood. Pluck the feathers while it's still warm. Hot water will loosen its feathers, except with waterfowl.

Cut the underside of the bird from neck to tail and remove the entrails. You can eat the heart and liver after cooking. Remove the head and feet, and butcher the rest. Do not skin birds in a survival situation, as doing so removes nutritional value.

Boil the meat for 20 minutes. You can also roast young, non-predatory birds.

You can use the feathers for insulation, tinder, and fishing. Anything else you don't eat makes good bait.

Related Chapters:

- Clubs
- Cord
- Cooking

REPTILES AND AMPHIBIANS

Reptiles are a good protein source and relatively easy to catch. They may have parasites, but you can eat them raw in an emergency.

Lizards

Catch small lizards by the tail and/or use a pit trap such as a mini below-ground solar still. Be careful of snakes that might follow them.

To prepare a lizard, start by removing its head. Make sure you cut down past the poison sacs if it has them. Cut open its skin from the anus to the neck. Pull out the internal organs and discard them. To remove the skin, roast it. When the skin splits, ease it off from the top (where you cut the lizard's head off) to the bottom. Roast or boil the meat.

Snakes

Some snakes are poisonous, but they're still edible. As a rule, stay away from snakes unless you're desperate for food or are certain they're not poisonous.

To catch a snake, use a forked stick to pin it down just behind its head then club the back of its head or stab it with a spear to kill it. Beware of it playing dead.

Chop its head off behind the stick (on the tail side) so you can flick it away. Bury it once it stops moving. Make sure you cut down past the poison sacs if it has any. Peel the skin off from head to tail and remove the entrails. Cut it into sections and boil or roast it. The meat will be rubbery, but edible.

Frogs

Frogs that are not brightly colored are edible, unless they have an "X" on their back. Do not eat or handle toads unless you know they're edible.

Find frogs near water, and then use a light to dazzle one and hit it with a club. Gut it and remove its skin, since it might be poisonous. Clean it well and boil or roast it.

Turtles and Tortoises

Most turtles and tortoises are edible after they're cooked, though the box turtle and the hawksbill sea turtle are not.

The box turtle is native to North America and Mexico. It has a domed shell. The shell pattern may be different than in this picture depending on the exact species.

Distinguish the hawksbill sea turtle from other sea turtles by its curved beak.

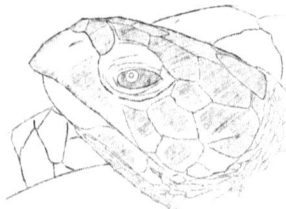

To prepare a turtle or tortoise for eating, first chop its head off. Use some bait to get it to stick his head out. Hang the turtle or tortoise upside down to let the blood drain out. Cut away the lower shell and cook the meat in a soup.

Related Chapters:

- Water Distillation

DEAD GAME

The large muscle areas of dead game that is not poisoned or decomposing are safe to eat. Check to make sure that it appears well-fed and is clean-smelling.

Do not eat it if it has:

- A bad odor before, during, or after boiling.
- Collapsed eyes.
- Discolored flesh.
- A sickly look.
- A slimy feel.

Cut the meat into small cubes and boil it for 30 minutes. Eat a little and wait 30 minutes to see if there are any ill effects. Most toxins act within 30 minutes.

FIRE

Building a fire while evading your enemy is risky. Only do it if it's absolutely necessary—that is, in cases of extreme cold, to purify water, and/or to cook food.

Better alternatives to fire-building are to:

- Build a shelter and insulate your body.
- Collect fresh water.
- Eat things you don't have to cook.

If you do decide to build a fire, conceal it as much as possible. Build it behind a natural barrier you put between you and your enemy. This will block the light, protect it from wind, and reflect heat back to you.

Consider the wind for safety and tactical reasons. It will flicker the flames and move smoke. Dusk, dawn, or bad weather will keep the smoke down. Smoke rises straight up on a fine day.

For safety reasons, never leave a fire unattended.

GATHER FUEL

Gathering fuel is the first step in building a fire. There are three types of fuel: tinder, kindling, and main fuel.

Gather the amount you need to last for the duration of the fire. Stack it close enough that it's handy, but far enough from the flames to be safe.

Dead wood is best since live wood contains a lot of moisture. Do not use wood from a biologically contaminated area, and protect all fuel from moisture.

To find dry fuel in wet weather, look:

- Inside hollow logs.
- For wood that is off the ground.
- For wood that stands vertically.
- Under the first layer or two of tree bark.
- Under the snow.
- Under top layers of foliage.

Tinder

Tinder is any material that takes only a spark to ignite. It must be very dry. Here are some examples:

- Birch bark.
- Char cloth.
- Cotton fluff.
- Down.
- Grass.
- Powdered fungi.
- Shaved bamboo.
- Shredded plastic or rubber.
- Termite nests.
- Thread.

- Toothpick or smaller-sized twigs.
- Waxed paper.
- Wood dust or shavings.

You can get tinder to burn better by saturating it with petroleum products (Vaseline, Chapstick, hand sanitizer, insect repellant, gas, etc.). You can also do this to the kindling.

Char Cloth

Char cloth is like a pre-made tinder. It's highly combustible, slow-burning, and easy to make. To create some:

Make a small hole (1mm max) in the top of a small, airtight tin.

Place (do not pack tightly or throw in) small squares of 100% natural cloth inside the tin. Any 100% natural cloth will work, but it must be 100%—an old cotton t-shirt, for example.

Put the tin in a gentle fire or in the embers. Do it so you will be able to see the smoke coming out of the hole in the tin. It may catch on fire. This is okay; it will burn itself out. When no more smoke comes out, take the tin out of the fire.

Let the tin cool down a bit before opening it, or it may ruin the cloth.

You want the cloth to be completely black, a little soft, but not too fragile. If it crumbles, it's overdone, and you'll have to start again. If there are brown patches, put it back in the fire for a little longer. Once it's done, separate each piece gently.

Tinder Nest

Constructing a tinder nest gives you a chance of starting a fire in a wilderness setting.

Collect lots of fluffy, stringy, fibrous materials, such as bark shavings, dried grass, or lichens. Shred it with your hands. Aim to make the strands as soft as possible.

Get the largest pieces and mold them into a palm-sized "bird's nest." Gather the next biggest pieces and add them to the middle of the nest. Do not pack them down. Repeat this process, using smaller and smaller fibers, until there's nothing left.

It is possible to ignite a tinder nest from sparks, but putting a char cloth and/or an ember inside it is better.

If the tinder will not ignite, it might be because:

- It's not dry enough.
- It's woven too tight.
- You're blowing too hard.
- You didn't knead it enough.

Kindling

Kindling is the material ignited by the tinder, which will burn long enough to ignite the fuel. You need a good supply of it and it must be dry.

Look for pencil-thick, soft wood. Resinous wood is good, because its sap is flammable. Feathering the wood by shaving shallow cuts into it will make it catch fire more easily. This is a feather stick.

Some non-natural products, such as a plastic spoon or a piece of a flip-flop, make good kindling because they catch fire easily and stay alight for a while.

Main Fuel

This is what keeps the fire going. It's okay if it's a little damp, but that will create more smoke.

Some good materials for main fuel are bamboo, coal, dry dung mixed with grass and leaves, dry peat, non-aromatic wood, etc. Hard woods burn longer. They are found on broad-leaf trees.

To split a log without an axe:

- Break the log over a rock.
- Burn it in the middle to weaken it, then stomp on it.
- Use a fork in a tree as a fulcrum. Put the log in in the middle of the fork to snap it.

TEEPEE FIRE

The teepee fire is a common fire lay that gets its name from its shape. It is fast to build and fast to burn, which is ideal for the evasive survivor. A teepee fire is mostly kindling, but you can add main fuel if you want it to burn longer.

Dig a depression to build the fire in. This helps to conceal the fire and block the wind. Even a small depression is better than nothing. Surround it with dry, non-porous rocks or other things to increase wind protection and block light.

For safety, clear a small area around the fire of anything flammable. Don't make the clearing too big otherwise it will be harder to cover it up when you're done. A radius of 1/2m (20in) is sufficient.

Place a bed of tinder (or your tinder nest) in the center of your firepit and ignite it.

Gently place more tinder on it, bit by bit. Be careful not to smother the flames. As the fire grows, start to lay kindling over it piece by piece to create a teepee. Continue to add larger and larger pieces of kindling , continuing the teepee shape.

Once your fire is strong enough, add pieces of main fuel. Again, be careful not to smother it.

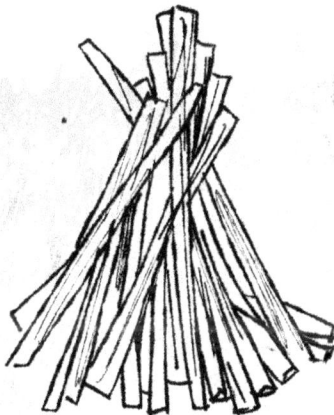

If you are unable to start a fire in a pit, it may be because of a lack of airflow. Try building it on a small mound of dirt instead, though this is not a good idea in tactical situations.

When the ground is wet or snowy, build a platform from green logs or dry, non-porous stones.

Another useful type of fire for when evasiveness is not an issue is the pyramid fire. A pyramid fire gives of a lot of heat and creates a big light signature, which is ideal for rescue. It's also good for cooking.

Related Chapters:

- Cooking
- Igniting the Fire

DAKOTA FIRE HOLE

Building a fire inside a Dakota fire hole has many advantages for the evasive survivor. This kind of fire is:

- Easier to light in high winds.
- Easy to cook on (lay green sticks over the hole as a grill or cook on the coals).
- Energy-efficient (requires less wood).
- Fast to kill and conceal.
- Less smoky.
- Reduced light signature.
- Able to be turned into a modified fire bed.

The downsides are that it is labor-intensive (you need to dig), and it doesn't radiate much heat.

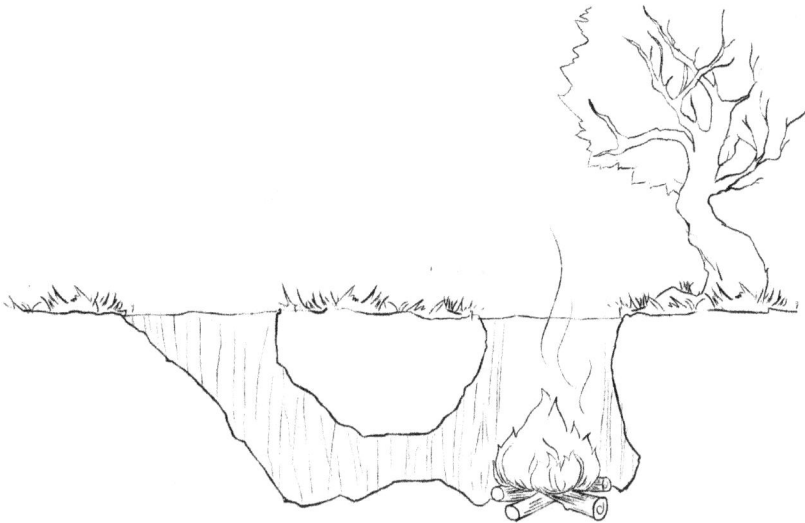

Try to find dry, compact soil to dig in, preferably away from large roots and rocks. A spot under a canopy of leafy foliage is ideal, as this will disperse the smoke and help to keep the rain out.

Dig a hole in the ground 30cm (10in) deep and 20cm (8in) wide. This is the fire hole.

Create a ventilation shaft on the upwind side of the fire hole. Make it 30cm (10in) away from the fire hole and 30cm (10in) wide. Angle the shaft so it connects to the bottom of your fire hole.

Create a trench around the hole to divert any water in case of rain.

Build your fire in the fire hole.

Related Chapters:

- Improvised Beds

IGNITING THE FIRE

Igniting the fire is the hardest part of fire-building, especially if you don't have matches or a lighter.

Light the fire from the upwind side, so the wind blows the flames into the rest of its structure. If it's windy, use your body to block the wind until the fire's going strong.

If you have matches, do everything you can to conserve them.

- Dry a wet match with static electricity. Roll it in some dry hair. Put pressure on the head when striking.
- Light fires without them whenever possible.
- Split each one in half lengthwise so you have two.

Using Sparks and Char to Start a Fire

Putting a spark directly to tinder to start a fire is possible, but difficult. Directing sparks into a tinder nest is easier, and directing them onto char cloth inside a tinder nest is the easiest. To do this:

- Place your char cloth inside a tinder nest.
- Direct sparks onto the char cloth. It will start to smolder.
- Gently fold the sides of the tinder nest to touch the smoldering char cloth.
- Blow on it to set your nest on fire. The redder it gets, the harder you can blow.
- Use this burning tinder nest to start your fire by putting it under your kindling.

Survival Fire Starters

There are many different types of survival fire starters on the market, but they are all essentially one of three types: Ferrocerium, magnesium, or flint and steel.

Ferrocerium is the most versatile. You don't need a special striker and it is the easiest to get a spark from. However, you can start a fire with any of these using basically the same method. If you buy one, it may come with instructions, which you should follow. If not, here's the general method:

- Place the ferrocerium rod (or whatever it is) onto your char cloth (or tinder nest) at a 45-degree angle.
- Put your scraping tool at the top of the rod and hold it there firmly at a 45-degree angle to the rod.
- Pull the rod back using a slow and steady motion. It is important to pull the rod back as opposed to moving the scraper forward, as this gives you more control and a better rain of sparks.
- Keep applying pressure as you pull the rod back so you get more sparks.
- The sparks will fly down onto your char cloth/tinder nest.

Ferrocerium doesn't require a specific scraper to use it. The back of your knife (never use the sharp side), hard rocks, hard bones, broken glass, and other things will work.

Batteries and Metal

Create sparks with a battery and anything metal by touching the metal on the positive and negative terminals of the same battery at the same time.

Do not touch the bare metal while it is touching the battery terminals.

Here are some examples that you can build on depending on what you have:

- Connect a piece of aluminum foil to both sides of an AA battery.
- Place a metal knife against both terminals of a cellphone

battery. This is not recommended unless you have no other use for your cellphone.

- Using wire (preferably insulated), attach one wire to the positive and the other to the negative terminal. Put both the loose wire ends over your tinder and touch them together to create sparks.
- Gently rub the terminals of a 9-volt battery onto steel wool.

Rocks

You can create sparks by hitting two rocks together, but the rocks must have specific characteristics. If you're not familiar with the rocks in the area, you'll need to test them, which will create noise.

Rock A must be pyrophoric, meaning it has to be something that ignites spontaneously in air at or below 55C (130F).

Iron is a pyrophoric and it's in steel, which is why the flint and steel method is so common. Carbon steel is the best. Other steels may not work well. Stainless steel, for example, is too hard.

In the wild, you can use iron pyrite, or fool's gold. This is often found in the same locations as chert (rock B), inside sedimentary rocks such as coal, limestone, shale, etc.

Iron pyrite looks like gold, but is lighter in color. It may look dull or tarnished or have greenish-black streaks.

Another substance you could use is marcasite. Marcasite has similar characteristics to pyrite, but isn't as shiny. Fresh surfaces are a pale to very pale yellow and have a bright metallic look. It turns brown with a black streak when tarnished.

Neither pyrite nor marcasite can be scratched with a knife.

Rock B must be very hard and preferably sharp when fractured.

Chert or flint is ideal, and is commonly found within other sedimentary rocks (a rock that appears to be in layers) such as chalks and limestones. Look in/near water sources (lakes, riverbeds, etc.).

Find a rock that you think may contain flint/chert and check inside it by hitting it with a medium-sized round, hard rock. Strike down on the slim edge of the "flint rock" at a 90-degree angle.

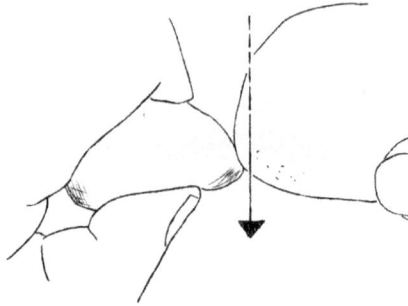

The color of flint/chert varies. It may be white, gray, black, deep red, deep blue, or another color.

Other rocks that work include agate, jaspers, quartz, and more. Look for rocks that have a non-porous, smooth interior with a glassy/milky/waxy appearance. Experiment to find what works best.

Once you've found a good rock, you need to create a sharp edge, unless it has one already. Use the same method as you would when creating a stone knife.

When you have your two rocks (A and B) ready, use rock B to strike down with a glancing blow on rock A. Rain sparks on your char cloth and/or tinder nest.

Focusing the Sun

To create a fire with the sun, you need a sunny day, but doing so is not ideal, since the smoke will send a clear signal to your enemy.

To use this method, focus sun-rays through a clear, curved, lens to form the smallest spot of light you can.

Eye glasses, a clear container filled with water (plastic bag, condom, water bottle, etc.), a magnifying glass, a piece of a clear glass bottle, etc. all make good lenses.

Focus this spot on your tinder nest and blow gently as it glows.

Chemicals

Mixing specific chemicals together will create a reaction to start a fire. Grind them with rocks or put them in the friction point of any friction fire method. Handle them carefully and keep them off metal.

Use:

- Three parts potassium chlorite (throat tablets) and one part sugar.
- Nine parts potassium permanganate (skin condition medication) and one part sugar.
- Three parts sodium chlorate (weed killer) and one part sugar.

As a last resort, use any pyrotechnics, such as flares.

Lighting Gas, Oil, or Fats

Consider carefully before lighting gas, oil, or fats, as they are highly flammable. Never light them directly. Use something like a feather stick instead.

Keep any extra supplies at a safe distance from the fire, and be careful not to get any of the fuel on your skin, especially in cold weather.

To burn oil and water, use two parts oil with one part water. Use a barrier, such as a metal sheet, to stop it soaking into the earth.

To burn gas, mix it with sand. When using it as a quick fire-igniter, put a gas-soaked rag in the tinder and pour more gas over the kindling. Wait a few seconds, then light the rag. If the fire doesn't light the first time, check for sparks or embers before adding extra gas.

Related Chapters:

- Knives
- Gather Fuel
- Friction Fires

FRICTION FIRES

A friction fire is one you make by rubbing pieces of wood together to create heat. Producing enough heat will produce an ember, which you can then use to ignite your tinder.

There are several ways to make a friction fire. None are easy, and they all need practice.

All friction fires have two main wooden components: a fireboard and a drill. The type of wood you use for these components will determine your chance of success.

Find the driest dead wood that is standing upright and use the same species, and preferably off the same piece of wood, for both components. Using different woods is possible if that's all you have, but it will be harder.

You want wood that's not too hard or too soft. If it's too hard, you won't be able to create dust for the ember, but if it's too soft, it will break under the pressure. An easy test is to press your thumbnail into the wood (not the bark). If you can leave an impression in it without bending your thumbnail, it should work.

Do research depending on your location. For example, in temperate climates (such as North America), alder, poplar, and willow are known to be good choices.

Once you find something that works, carry it with you and/or note it down.

The Fire Plough

The fire plough is not the easiest way to start a friction fire, but it doesn't require any tools to construct.

In this description, the fireboard component is the base, and the drill is the plough.

Unlike other friction-fire methods, this one will be easier to use if your fire-board is a little softer than your plough.

For the fireboard you need a piece of wood at least 5cm (2in) wide.

For the plough, find a tapered stick 25cm (10in) long. Make the thin side (the tip) about 1cm (1/2in) in diameter by rubbing it against an abrasive rock. If the tip is too big, it won't create enough heat. If it's too small, it will dig into the fireboard, making it harder to move back and forth.

Now that you have your components, make a groove in your fire-board. Find a position you can sustain for a while and hold your fireboard firmly so it will not move.

Hold your plough perpendicular to the fireboard. Place one hand 2.5cm (1in) above the tip, push down on the end with the other hand. Push back and forth (plough) to create a groove in the fire-board 15cm (6in) long.

Once the groove is there, lower the end of the plough so that there is maximum contact between the two pieces of wood, while still being able to move it back and forth.

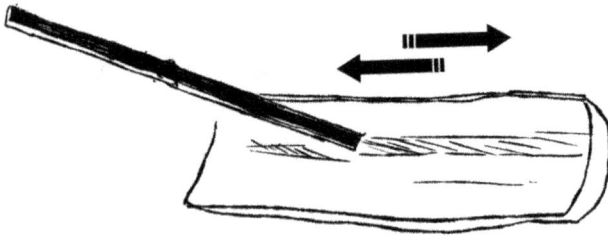

Plough steadily, using all of the groove. Experiment with pressure and speed until you get a good rhythm that produces smoke. A lip may form in the groove. Either push through it or stop before you touch it.

Once you start to see thick smoke and black dust, raise the plough a little to concentrate heat into the tip. Be careful not to destroy your dust pile, but at the same time, you want to touch it every other stroke.

Get over the fireboard and use your large muscle groups to plough. It takes effort to do this!

Accumulated dust will shorten your groove. Keep it at least 8cm (3in) long so you can maintain speed. Extend the groove towards you as you plough, if necessary.

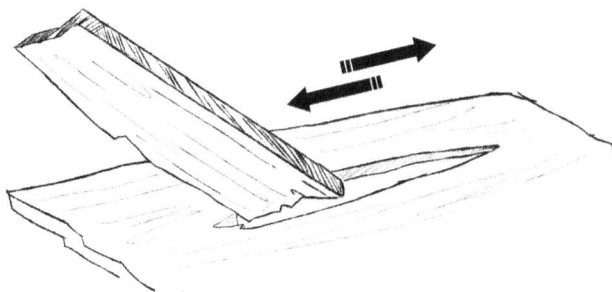

Eventually you will create an ember, which you will use to start your fire. Be careful not to destroy it.

Troubleshooting the Fire Plough

The finer the dust you make, the better. If you're creating dust that's too big, use less pressure and more speed.

If the groove is deepening too fast:

- Lower the end of the plough.
- Use less pressure and more speed.
- Do a combination of the above.

If a black, shiny glaze forms, you will lose friction. To avoid this:

- Clean the glaze off with a rock.
- Drop a tiny bit of sand in the groove.

If it gets too hard to plough, your groove may be getting too deep. Make it wider (as opposed to deeper) by shifting pressure 45 degrees to the side of the groove.

Bamboo Fire Plough

This is a fire plough made from bamboo.

Cut off a section of bamboo about 1/2meter (2ft) long. Split it in half length-wise, then create a small hole in the center of one of the halves. Place the piece of bamboo with the hole in it over some tinder. The curved side faces up and the tinder is directly underneath the hole.

Place the edge of the other half of the bamboo into the hole. Apply a little downward pressure as you plough.

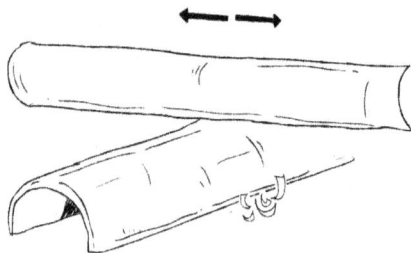

Bow-and-Drill

The bow-and-drill technique is the most efficient way of starting a friction fire, but it takes the most effort and resources to set up.

There are five parts to this tool:

- Bow and string.
- Drill.
- Socket.
- Fireboard.
- Ember patch.

The drill must be as straight as possible. If you have the resources to do so, carve it.

Make it 2.5cm (1in) in diameter and 25cm (10in) long. Shape both ends into blunt points, making one smaller than the other. Do not taper it.

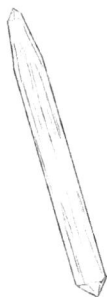

The fireboard is 2.5cm (1in) thick, with a flat bottom. Carve a small depression 2.5cm (1in) from the edge of one side. A sharp rock can do this.

For the bow, find a resilient green stick about 2.5cm (1in) in diameter and a little longer than one of your arms. It must be light, stiff, and strong, with only a little flex.

The bow string can be made of any cord. Paracord or shoelace is good. Improvised cord makes the job harder, but is okay if you have nothing else. Tie the cord from one end of the bow to the other. Make it taut.

Use the socket to push down on the drill. Find the hardest green wood you can—you need the moisture for lubrication. Get a piece that fits nicely in your hand and has a slight depression on one side.

The idea behind this method is that ember patch will catch and transfer the ember to your tinder. Dry bark works well for this.

Once you have all your pieces, wind the bowstring around drill once, so that the drill is on the outside of the cord (not in between the cord and the bow, in other words). Make sure the drill does not slip in the cord by trying to slide it up and down. If it does slip, tighten the cord on the bow.

Put one foot near the depression on the fireboard to keep the board steady. Rest your other knee on the ground.

Spit into the depression of the socket for lubrication.

Place the small end of the drill into the depression of the fireboard. Put the socket on top of the drill and apply a slight downward pressure.

With a smooth sawing motion, move the bow back and forth to spin the drill. Use the full length of the cord. Keep the drill vertical and saw evenly.

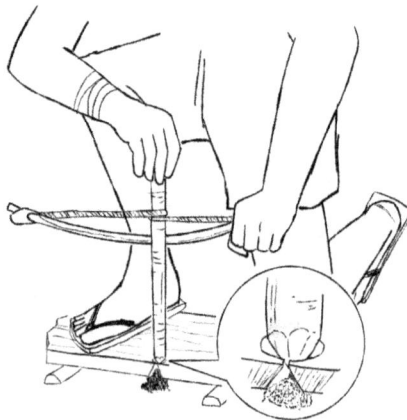

Start steady and gradually increase speed. Do not go too fast yet.

Drill enough to make a depression in the fireboard and the bottom of the drill black. This is called "burning in the set."

Carve a notch from the edge of the fireboard like a 45-degree slice of pie angling out from the center of the burnt-in depression. Do it on the side facing you as you drill. Ensure it's not too small and not too big.

It is important that you burn in the set before carving the notch.

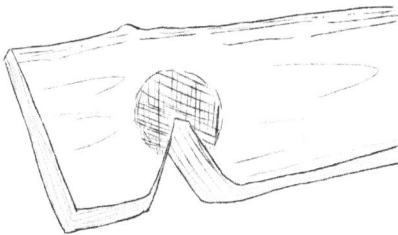

When you're ready to start a fire, place the ember patch under the notch in the fireboard.

Start drilling as previously described.

When smoke appears, apply more pressure and speed. Heavy smoke accumulating around the depression indicates an ember forming. Do at least 10 more full strokes.

Carefully tilt the fireboard away from you while holding the ember on the ember patch with a small twig. Gently tap the fireboard to ensure all the ember has fallen out of the notch and is lying on the ember patch. Remove the fire board.

Use the ember to create the fire.

Troubleshooting the Bow-and-Drill Method

If the drill won't stay in the depression:

- Apply more downward pressure.
- Increase the width/depth of the depression.

- Do a combination of the above.

If the drill won't twirl:

- Apply less downward pressure.
- Tighten the bow string.
- Do a combination of the above.

If the socket begins smoking:

- Apply less downward pressure.
- The wood is too soft. Lubricating it with something like animal fat or oil will help.

If there's no smoke and/or the bowstring runs up and down the drill:

- You're not keeping the drill straight. Lock your socket hand against your shin whilst sawing.
- Your drill is not carved straight. Fix it or get a new one.

If there's smoke, but no ember:

- Check that the pie slice notch is cut into the center of the depression.

Hand Drill

The hand drill is like a bow and drill, but you have no bow and no socket. This makes it harder but not impossible.

Make the fireboard, drill, and ember patch in the same as you would for the bow and drill method.

Roll the drill in the palm of your hands, running your hands down as you press it into the depression.

Starting a Fire with an Ember

Once you've created an ember, you must be very careful with it. Protect it from moisture, wind, etc. Shield it with your hands.

If needed, gently fan the ember with your free hand until it starts to glow. Once it's glowing (not before), carefully drop it into your tinder nest.

Gently squeeze the tinder nest around the ember so there is maximum contact between the tinder nest and the ember, but don't crush the ember.

Blow firmly, but not too hard. The ember will grow and start to smolder. Apply more pressure and blow more. Thick smoke will appear, and then the nest will burst into flames. Invert your tinder nest so the flames can burn up it.

Place the ignited tinder nest on the tinder in your fire. If you don't have a tinder nest, put the ember on whatever tinder you have and blow on it lightly until it creates flames. Add more tinder to grow the flames and build up your fire.

Related Chapters:

- Cord
- Gather Fuel
- Teepee Fire

FIRE MAINTENANCE

You have to nurse your fire until it's strong.

Gradually add larger bits of fuel, but be careful not to smother it. Once you have at least one piece of main fuel burning steadily, you can relax.

If you want to keep it going for a while, lay larger pieces of fuel on top of the flames in a crisscross method.

As an evasive survivor, you want to keep the flames low. Let the teepee fire collapse onto itself. You can force it down, but be careful not to smother it.

Extinguishing a Fire

Only keep a fire for as long as needed. Once you are done, extinguish it and camouflage the area.

If you built the fire inside a Dakota fire hole, fill in the hole and make the area blend in.

For a fire pit, flatten and douse it with water. Stir it to ensure it's completely out. If water is not available, use dirt.

Throw larger pieces of charred wood away in various directions, but not towards your shelter. Bury what you can of the smaller remains in your hole if you had one. Scatter the rest of the remains and make the area blend in.

If you have something to hold it in, you can carry some charcoal to make lighting the next fire easier, but you'll be carrying the smell of it too.

In a non-tactical situation, you can soft-kill a fire overnight by covering the embers with ashes and dry earth.

They will still be smoldering in the morning, and the fire will be easily to restart by adding dry leaves and blowing on it. Ensure you clear the area around it well, so it doesn't reignite on its own.

Related Chapters:

- Camouflage
- Teepee Fire
- Dakota Fire Hole

RESCUE SIGNALS

Setting up rescue signals is risky for the evasive survivor. It may lead your enemies to you, but there are some circumstances where you may want to try it anyway—for example, you're injured or sick and will die if you're not rescued. In this case, set up as many signaling devices as you can.

Note that there are ways to signal for rescue without alerting your enemy.

TYPES OF RESCUE SIGNALS

In a covert situation, don't use signals that your enemy can also see unless you're confident you will be rescued before they will capture you.

Avoid placing signals in areas frequented by people, such as roads, trails, man-made structures, inhabited areas, etc.

Create a signal and then hide in a place where you can observe it and anyone approaching it. Have an escape route if your enemy does come first.

In non-covert situations, set up as many rescue signals as you can. Stay with your vehicle, if you have one, since it's easier to spot. If you're on the move, keep to established trails and other places people and rescuers are likely to frequent.

SOS

SOS is the universal sign of distress. Its pattern is:

... - - - ...

This can be transmitted as:

- short short short
- long long long
- short short short

Audio

You can communicate SOS via tapping (tapping something metal on a pipe, for example) by using the length of pauses between the taps.

- Tap tap tap. Pause.
- Tap. Pause. Tap. Pause. Tap. Pause.

- Tap tap tap. Pause.

Anything that makes sound is good when you need rescuing. Trip fire alarms, break shop windows, or honk horns.

Radios

Radio signals are transmitted further when you're in the open, but you'll be more visible to your enemy that way.

Mayday is an internationally recognized radio distress signal. To use it:

- Press and hold the radio call button.
- Speak steady and clear.
- Wait for one second.
- Say "Mayday, Mayday."
- Give your location, preferably in map coordinates.
- Request immediate dispatch of emergency services.
- Release the call button and wait for a response.

A lack of response doesn't mean no one hears you. Repeat the information three times in a row before switching the radio off to conserve battery.

Repeat this at regular intervals.

Once contact is made, maintain communication until rescued. Give as much extra information as possible. List your injuries, supplies, and dangers. Do as instructed.

Visual

Any flash of light makes a good visual signal. Switch lights on and off in the SOS pattern. For example, use:

- Three short (one-second) bursts of light.
- Three long (three-second) bursts of light.

- Three short (one-second) bursts of light.

You can also write SOS or HELP and display it in a window.

In a wilderness situation, write SOS on the ground for rescuers in aircraft to see. Try to put it in a place that can be seen in any direction, and as is as high up as you can go. Make it big and use materials that contrast against the background. Bring material from another area if you have to.

A large triangle is another international distress symbol.

Flares

Flares are internationally recognized as distress signals. If you have a choice of color, pick one that contrasts the most with the background. Red is a good option in most cases.

Read the instructions carefully before firing.

If you're in a dinghy, hold the flare over the side to prevent damage to your vessel.

Always fire a flare in front of the possible rescue craft, not after it has passed.

Sea Dye Markers

Sea dye markers are visible for up to 5km (3mi), on average. Don't use them in rough seas or fast-moving water. It's a waste.

You can also use them to color snow.

Conserve the ones you don't use by rewrapping them.

Signal Mirrors

You can use anything reflective, like a polished soda can, as a signal mirror. Polish metal with sand.

Keep your signal mirror handy for immediate use. Hang it around your neck, for instance. Make sure the reflective side is against your body, so that it won't give you away in a covert situation.

To use a signal mirror:

- Bring the mirror close to your eyes.
- Place your hand between you and the rescue vessel.
- Angle the mirror so it flashes onto your hand.
- Move your hand away.

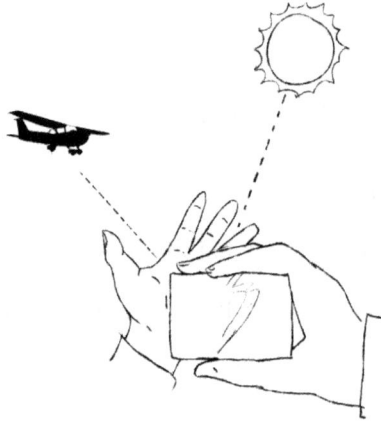

Don't direct the beam into an aircraft's cockpit for more than a few seconds. Stop once the pilot acknowledges you by dipping his wings and/or flashing his lights.

If you can't see the vessel, flash it in the direction you hear it coming from.

Even when there is no sign of a rescue craft, sweep the horizon with your signal mirror regularly in the opposite direction from where your enemy is.

Flags

Tying any piece of cloth (e.g., clothing) to a stick makes a good flag. The brighter and/or shinier it is, the better. Hold the flag to your left for dashes and to your right for dots. Use slightly longer pauses with dashes than dots.

Use figure-eights for exaggeration. Go to the left and make one, then go to the right and make another.

Fire

Establish signal fires in clearings and/or at high points, and down-wind of landing sites if possible. Keep them dry and full of tinder so they will light quickly when needed.

A single fire will attract attention, but three in a triangle is a definite sign of distress.

When vegetation is dense, you may need to create a fire on a lake or river for it to get seen. Build a raft to put it on and anchor it in position.

A bright fire is best at night, and a smoky fire is best during the day. Building a fire on metal makes it brighter, especially if the metal is polished.

Dark smoke stands out against snow or desert sand. Rubber or petroleum products (e.g., oil) will create dark smoke.

Light smoke stands out against dark forest. Use green vegetation and wet materials to create it.

Pyramid fires make good signal fires. To make one:

- Place two logs parallel to each other.
- Stack two more logs perpendicular to the first two.
- Continue to stack logs in this manner.
- Place tinder and kindling in the center.
- Cover the pyramid with smoke-producing material, if

applicable. This also protects the kindling from bad weather. Leave a gap so you can light it.

Torch Trees

Small, isolated trees make good pre-built signal fires. Place dry tinder in all the branches you can reach easily, and/or build fires at the bases of the trees.

On the Move

If you're leaving campsite permanently, leave information to help rescuers find you (assuming you're not being hunted). Include your:

- Travel direction. Ideally, use a big arrow, so it's visible by air.
- Departure date and time.
- Destination.
- Available supplies.
- Personal condition.

Continue to leave signs of your trajectory as you move. Bend things in the direction you're headed, for example.

THANKS FOR READING

Dear reader,

Thank you for reading *Evasive Wilderness Survival Techniques.*

If you enjoyed this book, please leave a review where you bought it. It helps more than most people think.

Don't forget your FREE book chapters!

You will also be among the first to know of FREE review copies, discount offers, bonus content, and more.

Go to:

https://offers.SFNonfictionBooks.com/Free-Chapters

Thanks again for your support.

REFERENCES

12PillarsOfSurvival.com. *Survival Stash*. 12PillarsOfSurvival.com.

Alton, J. (2016). *The Survival Medicine Handbook*. Doom and Bloom.

Auerbach, P. Constance, B Freer, L. (2018). *Field Guide to Wilderness Medicine*. Elsevier.

Chesbro, M. (2002). Wilderness Evasion. Paladin Press.

Department of Defense. (2011). *U.S. Army Survival Manual: FM 21-76*. CreateSpace Independent Publishing Platform.

DOD United States Department of Defense. (2011). *Survival, Evasion, and Recovery*. Pentagon Publishing.

Emerson, C. (2016). *100 Deadly Skills: Survival Edition*. Atria Books.

Fiedler, C. (2009). *The Complete Idiot's Guide to Natural Remedies*. Alpha.

Goodwin, L. (2014). *Prepping A to Z: Book A*.

Goodwin, L. (2014). *Prepping A to Z The Book Series Book B*.

Goodwin, L. (2014). *Prepping A to Z The Book Series Book C*.

Goodwin, L. (2014). *Prepping A to Z The Book Series Book D*.

Goodwin, L. (2014). *Prepping A to Z The Book Series Book E.*.

Goodwin, L. (2014). *Prepping A to Z The Book Series Book F*.

Hanson, J. (2015). *Spy Secrets That Can Save Your Life*. TarcherPerigee.

Hanson, J. (2018). *Survive Like a Spy*. TarcherPerigee.

Hawke, M. Hawke, R. (2018). *Family Survival Guide*. Skyhorse.

Lieberman, D. (2018). *Never Be Lied to Again*. St. Martin's Press.

Luther, D. *The Prepper's Workbook*.

Miller, T. (2012). *Beyond Collapse*. CreateSpace Independent Publishing Platform.

Morris, B. (2019). *The Green Beret Survival Guide*. Skyhorse.

Nobody, J. (2018). *The Prepper's Guide to Caches*. Prepper Press.

Terrill, B. Dierkers, G. (2005). *The Unofficial MacGyver How-To Handbook*. American International Press.

WA Police, SA. (2019). *Aids to Survival*.

Wiseman, J. (2015). *SAS Survival Guide*. William Collins.

United States Marine Corps. (2013). *United States Marine Corps Individual's Guide for Understanding and Surviving Terrorism*. United States Marine Corps.

US Marine Corps. *Kill or Get Killed*.

AUTHOR RECOMMENDATIONS

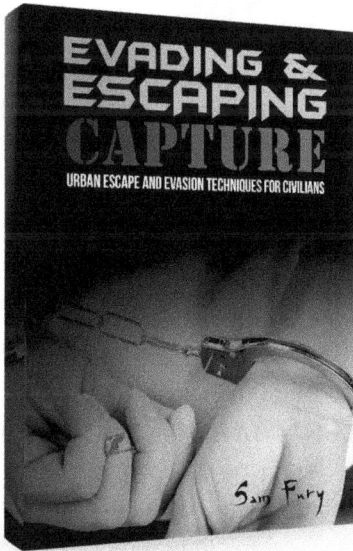

Teach Yourself Escape and Evasion Tactics

Discover the skills you need to evade and escape capture, because you never know when they will save your life.

Get it now.

www.SFNonfictionBooks/Evading-Escaping-Capture

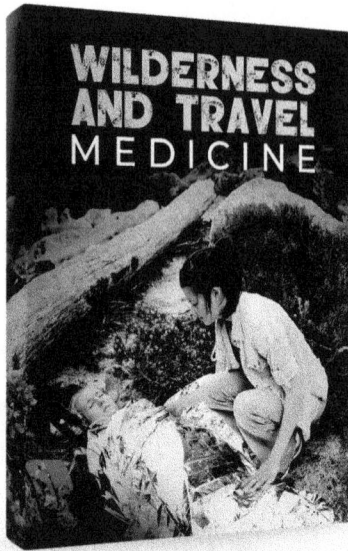

This is the Only Wilderness Medicine Book You Need

Discover what you need to heal yourself, because a little knowledge goes a long way.

Get it now.

www.SFNonfictionBooks.com.com/Wilderness-Travel-Medicine

ABOUT SAM FURY

Sam Fury has had a passion for survival, evasion, resistance, and escape (SERE) training since he was a young boy growing up in Australia.

This led him to years of training and career experience in related subjects, including martial arts, military training, survival skills, outdoor sports, and sustainable living.

These days, Sam spends his time refining existing skills, gaining new skills, and sharing what he learns via the Survival Fitness Plan website.

www.SurvivalFitnessPlan.com

amazon.com/author/samfury

goodreads.com/SamFury

facebook.com/AuthorSamFury

instagram.com/AuthorSamFury

youtube.com/SurvivalFitnessPlan